THE CITY   ELECTRIC

DUKE UNIVERSITY PRESS   DURHAM AND LONDON   2022

INFRASTRUCTURE

AND INGENUITY

IN POSTSOCIALIST   TANZANIA

the   city
electric

Michael Degani

Designed by Aimee C. Harrison
Typeset in Portrait Text and Helvetica Neue
by Westchester Publishing Services

Library of Congress Cataloging-in-Publication Data
Names: Degani, Michael, [date] author.
Title: The city electric : infrastructure and ingenuity in postsocialist
Tanzania / Michael Degani.
Description: Durham : Duke University Press, 2022. | Includes bibliographical
references and index.
Identifiers: LCCN 2022006715 (print)
LCCN 2022006716 (ebook)
ISBN 9781478016502 (hardcover)
ISBN 9781478019145 (paperback)
ISBN 9781478023777 (ebook)
ISBN 9781478092810 (ebook other)
Subjects: LCSH: Tanzania Electric Supply Company. | Electric power
distribution—Tanzania—Dar es Salaam—Reliability. | Electric power transmission
— Tanzania—Dar es Salaam. | Electric utilities—Tanzania—Dar es Salaam. |
Electric utilities—Corrupt practices—Tanzania—Dar es Salaam. | Political
corruption—Tanzania—Dar es Salaam. | BISAC: SOCIAL SCIENCE /
Anthropology / Cultural & Social | HISTORY / Africa / East
Classification: LCC TK1193.T34 D443 2023 (print)
LCC TK1193.T34 (ebook)
DDC 621.31/2109678—dc23/eng/20220801
LC record available at https://lccn.loc.gov/2022006715
LC ebook record available at https://lccn.loc.gov/2022006716

Cover photograph by the author.

This book is freely available in an open access edition thanks to
TOME (Toward an Open Monograph Ecosystem)—a collaboration
of the Association of American Universities, the Association of University
Presses, and the Association of Research Libraries—

and the generous support of Johns Hopkins University. Learn
more at the TOME website, available at openmonographs.org.

TO NATALIE

contents

acknowledgments

Ukiona vyelea, vimeundwa. It is slightly overwhelming to think of all of the friends, colleagues, and institutions that had a hand in this book's creation. Thanks first and foremost to Elizabeth Ault at Duke University Press, who shepherded a wayward author with grace and clarity and believed in the project from the jump. Two anonymous reviewers gave the manuscript a real workout, and it is better for it. I remain humbled by their thoughtful engagement and incisive comments. Benjamin Kossak was ever responsive about technical details and cartographer Jake Coolidge lent his considerable talents to produce two lovely maps. A generous subvention from the TOME initiative at JHU has allowed this book to go Open Access; thank you to Robin N. Sinn and Dean Smith for seeing that process through. Susan Weiner offered a last round of crucial, eagle-eyed edits and improvements.

I began thinking about what would eventually become *The City Electric* in graduate school, and no small amount of credit goes to my MA advisor at the University of Florida, Brenda Chalfin, whose foundational seminars in political economy and the anthropology of the state were, for lack of a better word, thrilling. She has remained a great mentor over the years and an intellectual inspiration; I am always slightly awed by her power, range, and energy, and am grateful she allocated some of it to reading through parts of this book.

It was Dr. Chalfin who encouraged me to apply to the anthropology program at Yale and mentioned I might look up a new assistant professor, Mike McGovern. When I went to the ASA meetings in New York that year, I happened to spot Mike waiting for the light on the other side of the crosswalk. I sometimes wonder how different my life might have been had I not decided, more or less on instinct, to flag him down and introduce myself. I owe him a debt of gratitude for the excellent training and patient reads, all of which allowed me to trust his praise and take in his measured critiques. His work is still to my mind a model for what anthropology can be.

Yale was an excellent place to learn the craft, and Kamari Clarke, Bill Kelly, Shivi Sivaramakrishnan, Erik Harms, Sean Brotherton, Helen Siu, and Karen Nakamura all shaped the sensibilities of the pages that follow. Special mention needs to be made of Barney Bate, whose untimely passing has already produced an outpouring of grief and remembrance amongst his friends and colleagues. Barney's was a rare light, and it still hurts to have lost it. His generosity, and his seminars on language and the public sphere, though, remain with me. Doug Rogers was a font of knowledge on all things postsocialism, energy, and ethics. Jim Scott's engagement with my work was a highlight, and the Agrarian Studies colloquium was, as far as I was concerned, the most intellectually vibrant space on campus. In seminars and over drinks, my friends and fellow grad students were deep wells of spirited critique, debate, and commiseration: Josh Rubin, Radhika Govindrajan, Annie Claus, Nate Smith, Minhua Ling, Ryan Sayre, Jun Zhang, Adrienne Cohen, Amy Zhang, Aina Begim, Atreyee Majumdar, Maria Sidorkina, Tina Palivos, and my academic "big sister" Susanna Fioratta. In a writing group with Alison Gerber and others, Joe Klett casually suggested I read Michel Serres—thank you.

Fieldwork in Tanzania was conducted with the generous support of the Social Science Research Council, the National Science Foundation, and the Wenner Gren Foundation, as well as grants from the Yale MacMillan Center and Program in Agrarian Studies. Two summers of language training in 2007 and 2008 helped acclimatize me to TZ. To my teachers, Aldin Mutembei and Deo Ngonyani: *asanteni sana*. Mohammed Yunus Rafiq, then a brilliant graduate student at Yale (now a brilliant anthropologist), ended up showing me around Dar for a stint of preliminary fieldwork and his charisma was a sight to behold. The arguments in this book have developed over many conversations with Yunus, often in a ramshackle mix of Kiswahili and English, and it is to Yunus I turn to see if I have captured a little of the *ladha* of Bongo, and of the country's complicated socialist legacy. In Tanzania, Kulwa "Inspector" Msonga and Rashidi Salum Ally shared their lives, and their friendships keep me tied to Dar. I wish to thank "Mr. Njola" whose assistance allowed me to secure a traineeship with Tanesco, as well as the array of clerks, inspectors, electricians, and contractors who tolerated my eager, if sometimes clumsy, questions. Learning about their lives has been the great privilege of this research. Professors Abu Mvungi and Patrick Masanja kindly granted me institutional affiliation with the University of Dar es Salaam's Sociology and Anthropology Department and presenting my work-in-progress to their razor-sharp students was invigorating. Brian Cooksey was a wealth of information and sometimes documents. Peter Bofin's knowledge of the ins and outs of Tanzanian politics was astonishing, and I always looked forward to a dinner with him out on the town.

Generous fellowships from the American Council of Learned Societies were crucial in allowing me to develop what would eventually become the monograph. While writing I have benefited from conversations with a number of committed and talented Africanists: Claire Mercer, Baruani Mshale, Julie Archambault, Andy Ivaska, Claudia Gastrow, Omolade Adunbi, Irmelin Joelsson, Ramah McKay, Molly Margaretten, Amiel Bize, Basil Ibrahim, Scott Ross, Yala Kisukidi, Erin Dean, Filip De Boeck, Anne Lewinson, Lori Leonard, Todd Sanders, Louisa Lombard, Chambi Chachage, and, especially, Jim Brennan. I can only hope what follows approximates their own work in rigor and care. A conference organized by Leonie Newhouse and AbdouMaliq Simone in Göttingen on the East African corridor was particularly generative, as were AAA panels organized by Kristin Doughty and Julie Kleinman. Thank you to Bjørn Bertelsen and Ruy Blanes, my fellow co-organizers of the Afrotopia conference in Zanzibar in 2017, which cast all manner of new light. Thank you to Kristin Phillips, who read through the manuscript and offered a number of fine insights.

Overlapping with these colleagues was a group of enormously talented scholars of infrastructure, energy, and beyond. It has been a privilege to think with Nikhil Anand, Hannah Appel, Jatin Dua, Janell Rothenburg, Stephanie Rupp, John Tresch, Etienne Benson, Canay Özden-Schilling, Dominic Boyer, Cymene Howe, Peter Redfield, Jamie Cross, Veronica Jacome, Tanja Winther, Nicole Labruto, Towns Middleton, Brian Larkin, Christina Schwenkel, and Laura Wagner. Along with Stephanie Friede and others, Ashley Carse co-organized a workshop on "Infrastructural Worlds" at Duke University in 2014 that stands out in my mind as an exciting moment of interdisciplinary encounter, and I cannot thank him enough for his feedback on a chapter at a critical point. A series of conversations with Eduardo Kohn about how grids think, if at all, lives in the roots of this book. Reading Paul Kockelman's account of infrastructure and parasites blew my mind, and I'm still recovering. I thank him for a series of wonderful comments and encouragements.

Finally, I have had the benefit of joining a passionate and often formidably brilliant department at Johns Hopkins University. A departmental manuscript workshop in 2018 helped shape the book in important ways and my thanks to Naveeda Khan, Debbie Poole, Veena Das, Clara Han, Niloofar Haeri, Alessandro Angelini, and Tom Özden-Schilling for their guidance and insights. Special thanks in particular to Anand Pandian, who read multiple chapter drafts, offered perceptive feedback, and provided crucial guidance in getting the book to press. His generosity of spirit and boundless sense of anthropological possibility have been a true inspiration. Finally, I am forever grateful for the chance to have spent a few years in Baltimore with Jane Guyer, who has long been a hero of mine. Beyond the anthro department, The Africa Seminar with Sara Berry, Liz Thornberry, Jeanne-Marie Jackson, and Julia Cummiskey has vitalized my work, as have engagements with Bill Connolly, Jane Bennet, Lester Spence, and P. J. Brendese in political science.

Outside of academia, a few important people deserve mention. Garrett Burrell is a wonderful poet and friend who understood the urgency of ecological thought in art and philosophy way before I did; I'm grateful for the long, winding conversations that helped me see it too. And thank you to David Shapiro, veteran electrical safety inspector, author, journalist, and publisher of the wonderful newsletter *The Flexible Conduit*. David found some of my work and has since become a vital interlocutor, pushing me ethnographically on technics and code, and displaying a deep intellectual curiosity about his fellow electrical experts and their work.

Starting the first real draft of this book coincided with the birth of my son, Owen. For most of his life he has seen, and occasionally helped, me "type" on Saturday mornings or in odd snatches during the day. It has been grounding to watch him grow through the good and bad days of piecing this work together. Finally, thanks to my wife and partner Natalie, who made space for this book and its occasional burdens, and who from the beginning heard my voice. If it comes through here, it is because of her. Her book is next.

Leo kuna maandamano ya wizi wanataka waongezewe siku, badala ya siku za mwizi
40 sasa wanataka za mwizi ziwe 60 watumie sms wezi wezako 20, kwangu ilikosewa
kutumwa na kwasababu inakuhusu nikaona ujumbe usipotee bure ni bora nikutumie
mhusika

Today the
thieves are holding a protest to demand that their days be increased from forty
to sixty, and any thief should send this text to twenty of his colleagues to let them
know. (This was mistakenly delivered to me but since it concerns you I figured
I'd pass it along.) —SMS MESSAGE

introduction

ETHNOGRAPHY OF(F) THE GRID

One afternoon as I was walking in Dar es Salaam, my friend Simon, an electrical contractor, forwarded me the text message above. This was the last era
before smart phones in Dar es Salaam became ubiquitous, and I laughed as I
read the chain letter on my sturdy Nokia X2. I had just gotten off a crowded
daladala minibus and was trudging my way over to the *kijiwe* (grindstone),
a roadside stretch of dirt where he and other contractors and electricians
waited for work and shot the breeze. Dappled by the shade of a large tree, the
spot attracted all manner of vendors: young kids bearing platters of *korosho*
groundnuts and loosie cigarettes; farmers in sandals leaning on carts of sugarcane and cassava; women carrying buckets of oranges and mangos at their
hips. At its center, a woman named Henrietta presided over a rickety wooden
stand, selling phone vouchers and a dozen different English and Kiswahili dai-
lies. Cars and pedestrians formed a steady stream in front of us, with a small
tributary veering off toward the glass doors across the street—the entrance to
a municipal branch of the Tanzania Electric Supply Company Limited, or
Tanesco.

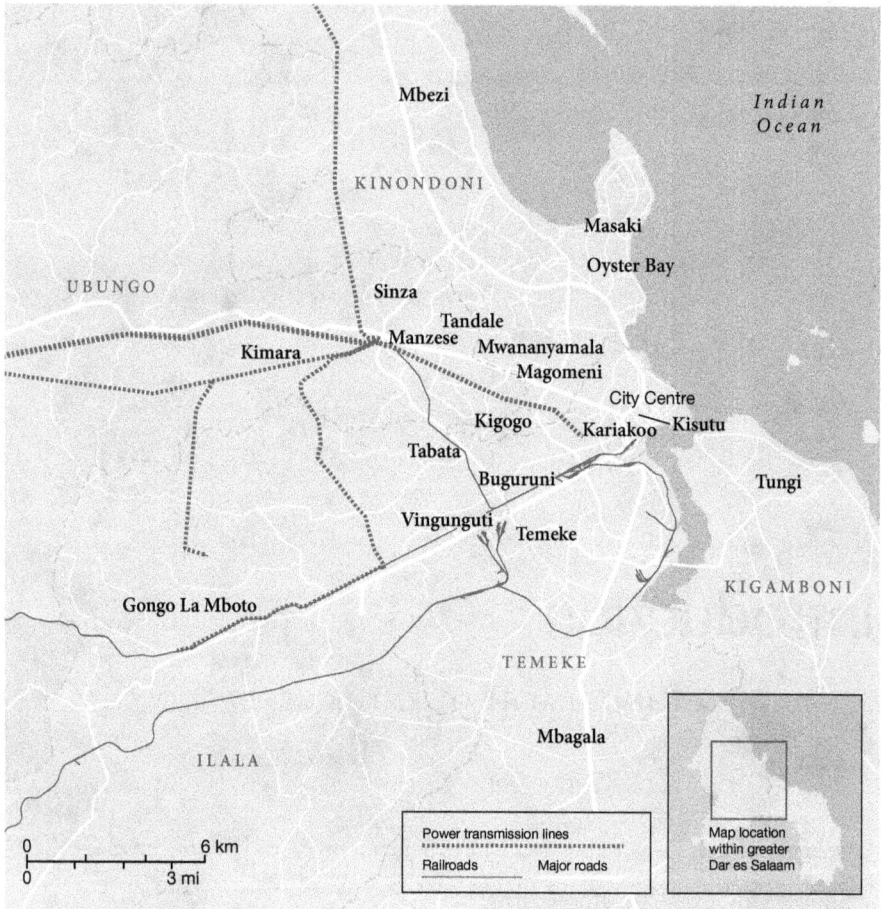

MAP I.1 Map of Dar es Salaam.

For Simon and his colleagues, dressed in collared shirts and dusty business shoes, the *kijiwe* offered a line of sight onto customers entering and exiting the building over the course of the day. Some looked beleaguered, perhaps angry about bills or broken equipment. Others clutched telltale oversized green folders holding applications for new "service lines" that would connect them to the grid. Should they cross the street, among Simon and his colleagues, they would find a pool of "drawers" (*vichoraji*) who, for a modest price, would provide properly formatted diagrams of their household wiring, endorsed by the stamp of a licensed contractor's seal. More conspiratorially, they might find electricians willing to repair or otherwise modify their existing service lines. Occasionally, a Tanesco employee would walk over to confer with a particular contractor, perhaps surreptitiously handing him a stack of green folders. The *kijiwe* was thus a kind of catchment area, a transitional zone where Tanesco's official operations shaded into the bustling informal economy that had grown up around it.

Such peripheral zones could be found thriving in the shadows of government buildings across the city. Along with some Tanesco branches, Revenue Authority and Immigration offices put up signs warning visitors to *Epuka Vishoka*, "avoid hatchets"—the popular term for agents and touts who falsely promised shortcuts for bureaucratic procedures. Officially, *vishoka* were considered thieves or parasites, figures to be avoided. There was a certain common sense to this. Over the course of my fieldwork, residents relayed numerous stories of being tricked out of their money (or foiling the would-be trickster). And yet *vishoka* were tolerated, even sought out. They provided valuable fixes to bureaucratic systems that were overlong, full of delays and, it was often alleged, full of their own scams. While not officially *of* the city's public infrastructure, *vishoka* were undeniably *in* it.

Wasn't Simon's text message saying as much? The text was a playful riff on the Kiswahili proverb *Siku za mwizi ni arboraini* (a thief gets forty days). "Forty" here is less a literal number than a period of spiritual trial and purification (e.g., forty days of mourning, forty days and nights in the wilderness). To say a thief gets "forty days" means that while he may be living the high life now, he can't outrun justice. From the standpoint of eternity, he has already been caught—his days are numbered. But the joke takes this stuffy piety and turns it into something more pragmatic and workaday. By demanding sixty days, Dar es Salaam's thieves are replacing a qualitative *category* of time (a thief *will always* get caught, but it might be after three days, seventy days, or in the afterlife) for a quantitative *length* of time. They reframe theft not as a wild antisocial eruption but a kind of reasonable work that they are entitled to pursue and whose terms may be negotiated.

With their blurring of theft and work, *vishoka* are symptomatic of the ambiguous state of the power sector in Tanzania and the broader political-economic shifts that have shaped it. Like many other developing nations in the post–Cold War 1990s, the once-socialist Tanzanian government embraced a host of neoliberal reforms designed to improve the efficiency and performance of its public institutions, including an ambitious plan to privatize its national power monopoly, Tanesco. But the situation quickly grew complicated as a series of political dealings and scandals derailed full privatization. Though Tanesco has pursued a number of austerity measures such as tariff increases and staff retrenchments, these policies have not translated to reinvestment in its degraded transmission and distribution network or its generation capacity. The result is that since the late 1990s, electricity has become more expensive and less reliable at the very moment that its practical and symbolic importance for life has increased—especially in Dar es Salaam, where nearly 80 percent of all connections to the national grid are concentrated. With economic liberalization came a flood of cheap electric and electronic imports—televisions, phones, refrigerators, fans, computers, blenders, and all manner of light industrial machinery—but the power to run them was supplied by an old, socialist-style parastatal entity seized with a new commercial spirit of cost recovery. A darkened TV screen, a warm fridge, a computer printout receipt for a service line never installed; "expectations of modernity" (Ferguson 1999) in the 2000s were not so much dashed as both renewed and deferred. *Vishoka* expressed this gap between ideal and reality—and have helped to bridge it.

Here, then, in the wake of a morally charged African socialism, this book explores how struggles over electricity became a key site for urban Tanzanians to enact, experience, and debate their social contract with the state. However, by "contract," I mean something more dynamic than this term usually implies.[1] For as power flows out through transmission and distribution lines to a bristling mass of consumers, and as payment flows back to Tanesco, a communicative circuit is formed, one that embodies a shared project of nation-building (*kujenga taifa*). And yet this circuit is fragile. Service interruptions, high tariffs, and aging materials all put pressure on the downward flow of current. In turn, consumer theft, nonpayment, or vandalism both accommodate these shifts and put pressure on the reciprocal up-flow of currency. Some of these perturbations are forgivably minor. Others reach a level of intensity that calls into question the reciprocal binding they presuppose. Indeed, at certain flashpoints, network breakdown threatens to erupt into arguments, violence, protests, or even regime change. This book zooms in on these moments to show how infrastructure becomes a site where the collective ambitions and commitments of an African postcolonial nation are both made and unmade, naturalized or suspended.

To do so, I turn to a tradition of anthropological work on money and exchange, as well as a broadly posthumanist style of thinking about networks and ecologies. The figure of the parasite will be important here, spanning as it does both Afro-socialist ideas about socioeconomic exploitation (Nyerere 1968; Brennan 2006a), as well as Michel Serres's cybernetic-thermodynamic insights into systems, signs, and signals (Serres 1982; Kockelman 2017). This redoubled parasite might be thought of as an exercise in "convivial scholarship" (Nyamnjoh 2017, 267), a lateral concept that, without erasing their differences, bridges certain "Western" and "African" intellectual traditions and highlights their elective affinities. As I elaborate, Serres's theory is built from the fables and picaresques of French agrarian life—a vernacular world of country and city mice feasting on scraps, of hares stealing into farmers' gardens.[2] Kiswahili epithets like *kupe* (tick) and *mnyonyaji* (a "sucker" of energy) are likewise rooted in a regional socioecological imaginary of collective resources diverted to self-ish ends, of ambiguous actors who are *in* yet not properly *of* the social body (Ngonyani 2002; Langwick 2007; Scotton 1965). Invoking these vernacular metaphors, and fusing them with socialist definitions of exploitation, the Tanzanian state vigorously curtailed the ability of party cadres and government bureaucrats to accumulate personal wealth, and policed ordinary people so that they would not stray into laziness, living off the vital flows—the sweat (*jasho*) or blood (*damu*)—of others (Brennan 2006a). Against visions of the post–Cold War era as a perpetually self-regulating End of History, I show how these old dramas still played out. The 2000s saw the purging of "parasitical" leaders who collected rents on national power generation and funneled them to their own private patronage networks, as well as "free-riding" consumers who stole power or didn't pay their debts. In some ways these social dramas were shaped by durable dispositions inculcated by the socialist period.

At the same time, these dramas of purification belied all sorts of interesting compromises, in which "parasites" were not just enemies to be expelled but uninvited guests to be (partially) tolerated. While a number of entrepreneurial politicians and civil servants exploited Tanesco's contracts with private companies for fuel, power, and supplies, only in the most egregious instances of state capture were any subject to anticorruption crackdowns. Meanwhile, as volatility and expense ramified down to the street level of the network, urban residents have contrived arrangements that deliver them power through theft or unofficial repairs. Tanesco responds by dispatching inspection teams—some of whom I accompanied—to monitor service lines and disconnect those that had been tampered with or had unsustainable debt. Patrols could unravel into fights with Tanesco consumers over the rights and responsibilities of citizenship in a postcolonial nation. But they also marked out spaces of tolerance that allowed some measure of unofficial or illegal diversion to continue.

In other words, this book will show the proliferation of "parasites" indexed growing misalignments between the overarching *social form* of the circuit and the *vital energy* that substantiated it (Mazzarella 2017)—that is, between the shared pact linking rulers and ruled and the distribution of material resources that made that pact plausible and efficacious, renewing citizen participation in it. Residents could choose to "play by the rules," enduring price increases and scarcity, thereby hewing to the form at the cost of its substance. Alternatively, they could acquire the substance in full—say by pirating current—but at the risk of contravening the very grounds by which that substance was channeled. As neither option was particularly appealing or workable for very long, politicians, residents, and electricians played with the alignment of energy and form—of electricity and the political compact its proper flow indexed—albeit within certain thresholds. They arrived at moral distinctions—between rent and plunder, rationing and "fake" rationing, debt and theft—that allowed some strain of electricity's economy of circulation while still preserving its basic integrity. These are exercises in *modal reasoning* (Degani 2017), and I will suggest they show us what it takes—and what it means—to sustain not just infrastructure but collective life more broadly under inauspicious conditions.

My analysis is based on approximately eighteen months of fieldwork in Dar es Salaam between 2011 and 2012, bounded by six months of preliminary research between 2009 and 2010, and six months of follow up between 2014 and 2018. This included time with electricians, contractors, consumers, Tanesco workers, managers, and bureaucrats. I utilized interviews, surveys, participant-observation, and archival and discourse analysis. Unsurprisingly enough, many of this book's insights and descriptions only came about after a long period of building trust and affection with different people involved in electricity's circulation. However, fieldwork also presented ethical challenges, particularly during utility patrols when residents pleaded, argued, cursed, and sometimes attempted to bribe the inspectors that cut off their power. There are colonial inflections to be found in almost any ethnographic project set in Africa (Assad 2006; Mafeje 1997), and the perverse privilege of witnessing the "infrastructural violence" (Rodgers and O'Neill 2012) of state disconnection brought those of my own to the fore—a privilege I work to unpack and historicize. In the end, rather than presume to be able to conduct fieldwork in a situation free of power relations, I too undertook a modal approach, defined by a threshold that I would not cross: namely, the active incrimination of anyone involved in power theft—inside Tanesco offices or out. Short of this, as I'll describe in the forthcoming chapters, I simply tried to move with situations in ways that preserved the project's ethical backbone.

My fieldwork was also enriched by interlocutors with no particular (or even literal) connection to electricity, but who connected me to the wider weave of urban life. Special mention should be made of Ally, who will appear in the pages ahead and whose friendship is one of the most enduring gifts of my time in Dar es Salaam. After living with his family and spending countless hours hanging out at his taxi stand, smoking cigarettes and drinking coffee, I consider him kin. Ally's biography, which spans a fraught Arab-African childhood, imprisonment for selling contraband cigarettes, work in a Shinyanga mine, a crippling leg injury, and a troubled but resilient marriage in Dar es Salaam, could serve as a Tanzanian outsider epic. For now I have simply cast him as an occasional, often dryly funny commentator on Dar es Salaam, a combination of Greek chorus and Muchona the Hornet (Turner 1967).

Over the course of four main chapters, *The City Electric* takes up electricity's economy of circulation as an ethnographic object and method. Roughly speaking, it "follows the circuit" (Marcus 1995) of current and currency to reveal distinctly postsocialist modes of power (1) generation, (2) transmission and distribution, (3) consumption and payment, and (4) maintenance/extension. Each chapter shows how, by degrees, actors divert, suspend, or restore—that is, modify—that stretch of the circuit. The conceptual upshot is to foreground the power "grid" as a living relation that waxes and wanes across a larger national ecology. We see how the pressures of multiparty competition strained power generation; how the resulting power cuts fed the suffering and resentment of an urban public; how attempts to bill consumers faltered on the illegibility of urban landscapes, and how Tanesco's own institutional thinning created a population of unofficial *vishoka* brokers. Like concepts, infrastructures don't have firm boundaries (Das 2015, 59; Star 1999; Larkin 2013); it is impossible to definitively mark where, say, the meter ends and the household begins, or where the power plant shades into the party. Caught within and constituted through these enmeshments, Tanzanians worked to distribute their crisscrossing forces and pressures so that, amidst the continuous unfolding of its politics, something like a national project might hold.

histories of/as infrastructure

An ethnography of infrastructural modification opens up useful perspectives onto the historical trajectories of neoliberalism, postsocialism, and African urban life. First, it offers a counterpoint to the many otherwise persuasive accounts of "Africa after the Cold War" (Piot 2010) whose cumulative effect is to emphasize a certain kind of formlessness.[3] At its most extreme we have the Mano River War that spanned Sierra Leone and Liberia, in which sociomoral distinctions between soldier and rebel, miner and fighter, cities and barracks—even war and peace—all seemed to dissolve into what Danny Hoffman, citing Gilles Deleuze and Felix Guattari, calls "production in general and without distinction" (2011, 104). Even peacetime megacities are likewise imagined as sites of constant flux and improvisation. Fantasies of evangelical ascension, transnational migration, or apocalyptic reckoning express the hollowing out of any meaningful trust in the "near future" (Guyer 2007), folding the End Times back onto the present to create a kind of eternal, protean now (De Boeck 2005; Melly 2017; Ferguson 2005a).

And yet from the vantage of the 2010s, the continuities—or at least subtler transformations—of continental experience are worth considering. If the liberal-democratic fantasies of the Washington Consensus have not come to pass, neither have the apocalyptic scenarios of unfettered capitalist dispossession—at least not everywhere. As James Ferguson has pointed out (2015), social welfare policies have only grown in South Africa, while strong state-driven developmentalist models (Mains 2019) across the continent suggest the emergence of a "Beijing Consensus" (Aminzade 2013, 268–72). In laying out these countertrends, the point is not to substitute one set of emphases for another, but rather to look closely at the complex imbrications of states, markets, aid, sovereignty, and neoliberalism (McKay 2017; Chalfin 2010), and the ways they resonate or not with popular aspirations or expectations of collective life.

It may be that formerly socialist African states provide an especially pronounced case where the sense of a shared social project lingers against the more centrifugal forces of competition and capture, even as in other moments those same forces are valued as spaces of entrepreneurial invention (Pitcher and Askew 2006). Mike McGovern (2017) has argued that states such as Guinea and Tanzania managed to inculcate a set of "durable dispositions" that provide ideals with which to critique capitalist accumulation and even resist the civil conflict it stokes. Thus, whereas Sierra Leone and Liberia ramped up into a regional war tied to control of natural resources, Guineans managed to avoid the export of that war across their borders. Similarly, many Tanzanians understand their own socialist path to have allowed them to avoid the rougher politics of their East African neighbors Uganda and Kenya, to say nothing of Rwanda or the Democratic Republic of Congo—at least for a time. This peace admittedly threatened to crack during the 2015–2021 presidency of John Magufuli, which saw a partial resurgence of socialist governance but also unprecedented violence and censorship—a development I discuss in the conclusion.

Socialism, as the saying goes, must be built. But how? McGovern emphasizes that in Guinea such durable dispositions were sedimented through a range of techniques: the iconoclastic destruction of indigenous paraphernalia, the reciprocal folklorization of "traditional" culture, and the widespread transmission of a revolutionary lexicon that wove itself into everyday speech and gesture (see also Yurchak 2006). Tanzania bore all of these features, including a particularly well-developed discourse on parasitism. This book explores how such an inheritance exerts a kind of drag on the present, frustrating the neoliberal tendency to depoliticize and "demoralize" the compact between state and citizen (Ferguson 2006, 71). And it does so by exploring a specific kind of technology by which this inheritance was sedimented: infrastructure.

"Study a city and neglect its sewers and power supplies (as many have), and you miss essential aspects of distributional justice and planning power" (Star 1999, 379). From pipes to canals to housing projects, infrastructure deeply influences the contours and textures of political community, and this makes it an incredibly rich site of ethnographic inquiry (Anand, Gupta, and Appel 2018; von Schnitzler 2016; Carse 2014; Fennell 2015). While the term infrastructure has no precise definition, it is generally associated with large sociotechnical systems whose scale and complexity embody decades and even centuries-long political dispensations (Carse 2016; Edwards 2003). The long modernizing century from roughly 1850 to 1960 saw the harnessing of massive amounts of force and energy into networked "spaces of flow" (Heidenreich 2009). This "integrated ideal" (Graham and Marvin 2001) presupposed a certain high modernist confidence in a rationally administered mass society and congealed an imperial fantasy of uninterrupted growth that only belatedly and briefly (ca. 1945–1979) included postcolonial African nations (see, e.g., Hoffman 2017).

The historical ascendance of neoliberalism is often thought to mark a radical disintegration of this ideal. By the post–Cold War 1990s, the ostensible "privatization of everything" (Watts 1994) was driven by an end-of-history triumphalism most fully realized in sales of state-owned assets such as factories and railroads, but also in the marketization of public provisioning systems such as heat and water. For critics, this seemed to herald the "death of the social" (Rose 1996), in which long-term commitments to universal service provision "splinter" (Graham and Marvin 2001) into variegated arrangements with specific communities and individuals who can pay or mobilize clientelist pressure as necessary (Anand 2017).

Ethnographic analyses, however, reveal a broader picture of change, continuity, and contestation. First, waves of reform often broke on the intransigence of infrastructural systems and their users. In the former Soviet Union, municipal heat remained flowing through industrial cities, not only due to "social norms in actors' heads" (Collier 2011, 213), but because the physical structure of its centralized system was designed to heat entire apartment floors at a go, making shutoffs of individual flats for nonpayment difficult. And even where technical systems have been redesigned, affective dispositions nourished by older forms often remain encoded, stubbornly, in bodily habitus. After relocating to mixed-income apartment complexes, for example, residents of

Chicago's South Side still longed for and sometimes cleverly rigged up systems to approximate the luxuriant "project heat" of their former public housing high-rises (Fennel 2015). Various projects to privatize or otherwise enervate public goods have in fact foundered on the dispositions embodied in and distributed across those systems, because what is at stake is not just vital flows but the kinds of social persons and worlds those flows create.[4]

More complicated still, as Stephen Collier (2011) has emphasized, is the fact that not all neoliberal reformers rejected the welfare presumptions embedded in public systems, but rather sought to make them more efficient. After all, many public systems in both the Eastern bloc and Global South had suffered fiscal crises and operational inefficiencies since the late 1970s and 1980s. Rather than gut or privatize them wholesale, reformers sought to improve them by incorporating certain elements of competition, price response, and so forth. This basically centrist tradition of neoliberal thought on regulation—what in chapter 1 I describe as a "politics of the air conditioner"—seeks to artfully tighten, loosen, or otherwise "nudge" individual users by reconfiguring an institution's "choice architecture" (Thaler and Sunstein 2008). By expanding the degrees of freedom at certain junctures and pruning them at others, one can "optimize" the social good (Özden-Schilling 2021).

Like such works, the chapters that follow open the black box of "neoliberalism" and show not only how it has unfolded on the ground, but how it sometimes surprisingly resonates with the social state it supposedly displaces. Power sector reform was a complex and contested assortment of specific policies, some reviled and some quite popular. When a private South African management company operated Tanesco for four years in 2002–2006, it unexpectedly won a high degree of approval by cutting off government ministries who weren't paying their power bills. This kind of "neoliberal populism" (Aminzade 2013, 256) highlighted the ways in which market reforms and good governance could converge with socialist discourses of morality and public service (Sanders 2008).[5]

Ultimately then, this book is not concerned to adjudicate the degree to which "neoliberalism" was unilaterally imposed or resisted. Its fundamental axis is not between state and market but between rulers and ruled, wealthy and poor, and the ways each manage to "take a cut" of the flows that tie them together in a collective ecology of circulation. As such, it seeks to put the anthropology of infrastructure in conversation with a range of scholarship that, one way or another, is concerned with the compromises and accommodations of "conviviality" (Nyamnjoh 2002, 2014), of "living together" (De Boeck and Baloji 2016). How do societies forge loyalty, govern exchange, or legitimate coercion when their founding premises seem less and less plausible (Mbembe 2001, 77)? How can the infelicities that beset the performance of any social relation be accommodated in ways such that its basic form holds? We might sense in these questions an ethical shading that leads to concepts of being-with-others (Das 2014), or to more politicized frameworks involving moral economies (Scott 1977) and public goods (Bear and Mathur 2015), or to the metaphysics of gift and sacrifice (Viveiros de Castro 2014). In any case, fifty years after independence, these are the questions posed by Tanzanians' everyday experiences with cryptic blackouts and private generators, unofficial repairs and surreptitious connections.

Finally, one might also sense in these questions a disciplinary bent that draws out a latent structuralism within the infrastructural (Rutherford 2016; cf. Pandian 2014). With its healthy suspicion of closure, a broadly actor network–inspired branch of infrastructure studies has cultivated an appetite for tracing out the proliferation of more-than-human assemblages (Latour 2008; Jensen and Morita 2017; Bennet 2005). These works emphasize that infrastructures are the marshaling of an unruly network of heterogeneous species and actants, and are thus constantly subject to "parasitic" resistance and interference, addition and "accretion" (Anand 2015). Despite some philosophical conundrums,[6] there is an important antisystemic thrust to this orientation, pointing as it does to diachrony, actuality, and what Charles Pierce called the "brute" secondness of events (see Kohn 2013, 91), and I have retained much inspiration from it. But as I elaborate below, a focus on modification takes a step back to ask about the *logic* of parasitic variation, of what *counts* as an intolerable breakdown or a sufficient repair in relation to a whole, however projected or virtual. Marveling at the Zimbabwe bush pump, its "fluid" ability to span shifting configurations of E. coli, materials, and users, Marianne de Laet and Annemarie Mol observe that "even if many of its elements are transformed, the whole does not necessarily fall apart" (2000, 247). How does one get a feel for this kind of emergent "whole" across variation? In the context of a power grid, I'll suggest, it will entail seeing infrastructure as a nested hierarchy of surfaces and depths, short-term and long-term cycles.

the gift of infrastructure

At any given moment, 900–1,500 megawatts of Tanzanian-generated electricity is dispersed to approximately 900,000 separate residences and businesses. Consumers send money back through digital prepaid or analog postpaid metering systems, and these payments are recorded at local branch offices and, ultimately, at Tanesco's national headquarters. In short, electrical current moves "down" to the decentralized, street-level branches of the network as currency moves "up" to the centralized offices (*ofisini*). And because electricity cannot be easily stored, this metabolic exchange must proceed indefinitely in an even, balanced rhythm. This rhythm is of course subject to multiple syncopations. Engineers work to redistribute supply to times and areas of high demand, while the state often subsidizes generation and maintenance costs for utilities that cannot sustain themselves on revenue collection alone. The basic move is to keep the movement of current and currency in continuous, reciprocal exchange, even if it must partially gamble on future repayment to do so.

One of the founding texts of Western anthropology was concerned with a continuous circuit of exchange: Bronislaw Malinowski's ([1922] 2002) analysis of the Kula Ring, in which armbands and bracelets flow in countercyclical directions across the Trobriand Islands, carving out channels through which ordinary sorts of barter can take place. This inaugurated "an anthropology of the gift" that moved through French ethnology (Mauss ([1925] 2016; Lévi-Strauss 1969; Lacan [1966] 2007; Bourdieu 1977) and, perhaps surprisingly, given the well-known animus of science and technology studies toward this tradition (Kockelman 2017, 47), forms the "genealogical backdrop to [Bruno] Latour's Network Theory" (Povinelli 2011b, 5). As gifts like Kula shells moved out into the world and then returned (whether in that same form or some other), they became reflexive media of sociality, creating a horizon of relatedness beyond that particular instant.[7] At its purest, speculates Jacques Lacan in his structuralist period, a gift is *nothing but* a token of that social relation, its superfluousness the point ([1966] 2007, 273; see also Zizek 2006, 11–12). Decorative armbands and shells may be "merely symbolic," but as such they have a ground-clearing efficacy: they are objects whose particular exchange performatively inaugurates the ability of two (or more) parties to exchange in general.

The structuralist linguist Roman Jakobson (1960) arrived at a parallel insight with the notion of phatic communication: acts that open, close, or otherwise configure channels through which we can exchange more substantive kinds of messages. The rub is that "making contact" is never merely technical, but is rather a symbolic exchange in its own right, freighted with social implication.[8] Jakobson notably drew inspiration from what Malinowski called "phatic communion" ([1923] 1972): "empty" formalities such as polite greetings or chatting about the weather that serve to bond the speakers more than convey referential information.[9] In a marvelous essay, Julia Elyachar (2010) analyzes the daily "phatic labor" of Cairene women who visit each other for tea to exchange pleasantries. These courtesies nourish relations that can deliver actual goods (favors, information, a word to the right person in government) should the occasion arise. Such visits are indeed a kind of maintenance work, ritually lingering upon and attendant to the relation as such. The homology to conventional infrastructure should be clear: pipes and wires do not always transmit water, power, or data—they are rather the conditions of possibility for those vital flows, the delimiting ground against which a substance/figure can emerge.

In Dar es Salaam, extending a service line is both a gift and a phatic communiqué. In an important sense it is "empty." Its installation does not determine how often or how much power a given household will consume. It simply opens a channel, elevating the household into a "long-term transactional order" with the state power company (Parry and Bloch 1989). As Tanja Winther aptly put it in her ethnography of electrification of Zanzibar, "the costly and solid physical connection between consumer and utility binds the two parties together like spouses in marriage: tied to each other through the good times and the bad times and with a high cost in the case of separation" (2008, 40).[10] And like any shared marital asset or communicative channel, a service line has its proscriptive constraints. While residents must pay for the parts and installation of the service lines, they do not actually own the wires, poles, or meters that comprise them. Electricians often spoke to me of the meter as a "gate" or border beyond which is Tanesco's property and responsibility; households cannot repair or alter service lines, nor take them if they move, nor certainly sell them. As Marcel Mauss famously theorized of the Gift ([1925] 2016), in a service line one doesn't just receive the object but something of the giver as well.

While I will do much to complicate this setup in the coming pages, I start with it here to emphasize that, at its most basic level, the power network binds a nested relation between consumers and utility and more broadly between citizens and the state. This relation is necessarily virtual. It has a social form (citizen/state) that can be inferred from its empirical manifestations (the circulation of current and currency). One might even picture a spectrum with virtual social form at one end and actual vital content on the other (see the current-currency circuit below). In the middle stands the service line. It is an empirically real artifact, costly and solid, given and received. But like phatic communication it is hollow in the middle. Its content is the form itself.

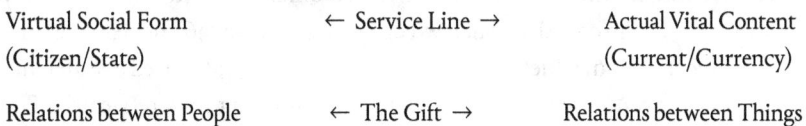

Virtual Social Form (Citizen/State)	← Service Line →	Actual Vital Content (Current/Currency)
Relations between People	← The Gift →	Relations between Things

Starting with form will allow us to understand the way politicians, consumers, and electricians have worked within it—or fallen out of it. We have, if you like, set up the rules of the system. But as we know from the Wittgensteinian tradition, to follow a rule is a complex affair involving the "interplay of obligation, coercion, and desire" (Das et al. 2014, 7). That the lived space delimited by such rules (or more accurately, the space from which we could infer such rules) will remain inhabitable or even intelligible is in no sense guaranteed, and often depends on the vagaries of history, as I'll show below. Form is a precarious achievement, always in danger of being "dismantled like a palace swept away upon the flood, whose parts, through the effect of currents and stagnant waters, obstacles and straits, come to be combined in a manner other than that intended by the architect" (Levi-Strauss 1966, 231).

socialist circulation

In the run up to independence in the late 1950s and 1960s, Africans entertained and debated different forms: national autonomy or regional federation (Wilder 2014)? Pan-Africanist unity or Marxist solidarity (McGovern 2012b)? Political scientist Jean Francois Bayart (1993) posited a spectrum of decolonization styles ranging from "conservative modernization" to "social revolution." Tanzania veered toward the latter, exemplified by first President Julius Nyerere's political philosophy of *ujamaa*. For Nyerere, *ujamaa* meant something like African socialism, but more literally translates to familyhood or kinship (Lal 2015). This was a variant on the regional politics of social (inter) dependency, in which a circuit of reciprocal rights and responsibilities structured the relations between master and slave, patron and client, or husband and wife (Glassman 1991; see also Eglash and Foster 2017). Dependency was a hierarchical but relational contract, with well-understood opportunities for social "exit or voice" (Hirschman 1970) should the more powerful not hold up their end of the bargain of patronage and protection (Ferguson 2015).

What pushed this patronage structure toward social revolution was *ujamaa*'s emphasis on collective sacrifice of both rulers and ruled in service of the future. Nyerere outlined the philosophy of *ujamaa* as early as 1962. The following year, in a bid to eradicate "tribalism" (*ukabila*) he stripped chiefs of executive powers they enjoyed under the colonial state. After a failed army mutiny in 1964, he constitutionalized one-party rule. In 1967 he delivered the Arusha Declaration, which nationalized industry and instituted a leadership code that prevented party leaders from accumulating private wealth. The cumulative effect of these policies was to flatten and unify national space under the party/state. Nyerere was revered as *Baba wa Taifa* (Father of the Nation), but his more specific honorific was *Mwalimu* (Teacher); a paternal authority, yes, but a strict one, a channel for something larger than himself. Where other newly independent African countries might take the easy way out by submitting themselves as Cold War client-states, or where the government might abandon its citizens for illicit self-enrichment, or where ethnic factions might capture and eat the national cake, both rulers and ruled were called on to forestall these temptations and instead sacrifice for the greater good. They were called on to bear the heavy hand of state discipline, to lend their energies and resources to anti-apartheid and anti-imperialist struggles beyond their own borders. And yet in exchange for withholding from themselves the selfishly parasitical pleasures of Western modernity, or capitalism, or ethnic particularism, they entered into a collective form: an orientation to a shared future in which Tanzania was self-sufficient, pan-Africanist, and free of exploitation.

While the English translation of *unyonyaji*—exploitation or parasitism—is likewise drawn from the philosophy of international socialism, it was also grounded in deeply local meanings. *Unyonyaji* literally translates to sucking (the verb form *kunyonya* could mean suckling at the breast) and refers to the diversion of vital flows—blood, food, money, health (White 1994; Weiss 1996; Myhre 2017)—that connected people in proper social relations. This was a pan-ethnic idiom grounded in the experience of both social and biological parasites and resonated with fears of blood-sucking vampires and witchcraft eradication campaigns that occasionally swept through colonial East Africa in the first half of the twentieth century (Brennan 2008; White 2000). Building on these associations, Nyerere's nationalist party (then known as Tanganyika African National Union [TANU]) promised, as a series of political cartoons viscerally depicted, to "cut the straws of exploitation" that had been inserted into the African body (Brennan 2006a, 399). Most immediately this referred to white colonial rule, as well as to the commercial dominance of "Asians" (*Wahinidi*) and "Arabs" (*Waarabu*) plugged into diasporic networks of trade. Animus against the latter culminated in the 1971 Land Acquisition Act, which expropriated and nationalized Dar es Salaam's urban housing stock, the majority of which was Asian-owned (Brennan 2012).

Eradicating parasitism thus resonated with popular structures of feeling, and this in turn helped justify or at least render plausible the state's escalating consolidation over political life. And yet ambiguities remained, not the least being that the postcolonial state exhibited strong continuities with its late colonial predecessor. While the first decade of independence coincided with a period of widespread economic growth, most African postcolonial states remained essentially nonindustrialized, with a narrow manufacturing base and a wide swath of low-margin peasant agricultural production marketed and sold by a state elite (Hart 1982). Most Tanzanians' standards of living were not appreciably transformed to resemble those of their former colonial masters. As during the colonial era, urban migration continued apace as people looked for commercial opportunities more appealing than rural life, swelling an awkward category of poor urban dwellers engaged in petty trade. Neither peasant (*mkulima*) nor worker (*mfanyakazi*), this essentially lumpen figure had a number of overlapping names: *muhuni* (thug), *kabwela* (the perpetually short-changed "little man"), or *mswahili* (a Swahili or "coastal" person). And, as in the colonial era, the state undertook moralizing campaigns of harassment and repatriation, seeing them as parasitic on the "productive" economy (Burton 2007). People's militias (*migambo*) and Youth League members (many of whom would otherwise be targets of such campaigns) conducted quasi-militarized "operations" designed to sweep out vendors, idlers, and other parasites who were deemed enemies (*maadui*) of the nation (Ivaska 2011; Brennan 2006b).

More subtly, Tanzanians were called upon to police themselves. *Naizisheni* ("nationalizers"), the African clerks and bureaucrats employed by the colonial state who emerged as party leaders, looked down on "Swahili" urban dwellers for not conforming to the proper discipline of modern national life. At the same time, they could not overlook the material inequalities between them. The result was a series of constant little exercises in combatting their own unwitting exploitations. Consider the following newspaper editorial from 1968:

> There are several types of unyonyaji. There is the person who exploits another person or the person who exploits the Nation. Although it is true that many of us really hate exploiters, but we always forget to ask ourselves if we ourselves exploit or not. Maybe you who have your radio, and every day at 600 you listen to music, world news etc., have you already cut your straw of unyonyaji regarding your radio? Have you already paid your radio license for this year? If not, why in your heart do you hate so much unyonyaji? Remember that the machines that broadcast the news every day is paid by the money of society. (quoted in Brennan 2006a, 396)

In other words: don't just direct your attention to the news, direct your attention to the channels themselves ("of society") through which it flows. This is indeed a kind of phatic communication, and it is not surprising that much of Tanzanian statecraft was devoted to similarly ritualized affirmations of order. As in Guinea (Cohen 2021), the new state invested considerable resources in cultural life, sponsoring dance troupes that "performed the nation" often through simplistic, over-formalized slogans paying tribute to Nyerere or the party (Askew 2002). Harassment of the unproductive poor mentioned above—never very effective—also had a cyclical, almost ritualized quality. In the absence of truly *collective* development, citizens were consistently relayed back to the form through which it might be eventually realized.

fossil developmentalism

The electric grid was an appealing extension of this centralizing, collectivizing logic. The idea that power networks could help create national subjects was a prominent theme in left-wing modernization theory, most famously expressed in Vladimir Lenin's maxim that Communism is Soviet power plus the electrification of the countryside (see Scott 2000, 166). Elizabeth Chaterjee (2020) calls this nation-building sensibility "fossil developmentalism," which she contrasts with fossil capitalism, wherein electricity is a mere technical input for extractive industry and money-making ventures. Unlike its enrollment in the project of fossil capitalism, postcolonial electrification was imbricated in "a shared moral project of developmentalism . . . partly decommodified and instead conceptualized as a national good and increasingly as an (imperfectly recognized) entitlement to be demanded from the state" (2020, 4).

During the colonial period of what was then called Tanganyika, electricity still very much operated in the fossil capitalist mode. Across Tanganyika and Kenya, small diesel generators and transmission lines owned by colonial enterprises traced out the extractive geography of railroad lines and telegraph cables that linked plantations and mines to port and sea. This extractive geography was thematized in widespread rumors of "vampires" (*mumiani*). Reputed to be Africans in the employ of the colonial state, *mumiani* were said to whisk victims into the back of motorcars, behind hospitals, or within other nefarious colonial enclosures, and then parasitically suck their blood via electrical wires and other industrialized implements (White 2000). In such rumors the parasite is not the African feeding off the productive colonial economy but rather the colonial state itself. In Dar es Salaam, too, electricity flowed mainly into white neighborhoods and government buildings, with some current reserved for public street lamps in the Asian-dominated market district.

After independence, the colonial power utilities were nationalized and the state embarked on a concerted effort to expand generation and extend access to its new citizenry. From the 1960s to 1990s it built a series of hydropower dams largely funded by Nordic social democracies interested in Tanzania's socialist experiment, and exemplifying its logic of centralized nation building (Hoag and Öhman 2008; cf. Tischler 2014). It achieved a 60 percent connection rate in Dar es Salaam by 2000 and a 30 percent rate in rural areas (van der Straeten 2015b). Consumers enjoyed low domestic tariffs facilitated by industrial cross-subsidies. For the thin stratum of mostly urban dwellers, power was reliable and affordable, despite and perhaps because it was only casually connected to the cost of actual generation and maintenance. While actual access itself was far from universal, the very fact of its ongoing extension functioned as a promise. As another socialist-era editorial proclaimed, "the march of pylons is a sign of progress."[11]

By the 1980s, the signs looked less promising. As part of a more general economic crisis, the power network began to experience high levels of transmission and distribution losses that could not be remediated by the postcolonial state budget nor by Tanesco, given the latter's "low self-financing ratios" (Gratwick and Eberhard 2008, 3950). In tandem with the broader package of liberalizing reforms, the World Bank proposed to finance its unbundling and eventual privatization. And yet what ensued in Tanzania and in power sectors was hardly a straightforward transition to market competition. Chapter 1 takes up this history in detail, but here it suffices to say that Tanesco exemplifies the "demise of the standard model" of privatization (Gratwick and Eberhard 2008). Its lynchpin was the introduction of so-called emergency power project (EPP) contracts, noncompetitive tenders with private companies (most of which had complex ties to state officials) for expensively produced, thermally generated power to make up for hydropower shortfalls. Chaining the power sector to these supposedly temporary private contracts had a deleterious effect. They locked in a downstream pattern of cyclical blackouts and emergency supply, while driving up debt and preventing investment in maintenance or network extension. The result was not so much a replacing of public infrastructure with private enterprise as their ambiguous blurring— the kind of "crony capitalism" familiar across the post–Cold War world, from Russia to the United States.

Hence, in the early 2000s, precisely when electrification was expanding in importance, its degraded infrastructure constituted a major moral and technical strain. Ordinary power consumers were left with higher tariffs, aging service lines, and sometimes were literally left sitting in the dark. In turn, the "power woes" became a key index of the state's fraying commitments, in particular the fraying of trust in the ruling party, which had in 1977 changed its name to CCM (Chama cha Mapinduzi, party of the revolution). To what extent did the collective premises of (fossil) developmentalism, embodied in the gift relations of infrastructure, still hold? To what extent had CCM's leaders, released from the moralizing strictures of a centralized party/state, become parasites, unduly eating at the expense of *wanyonywa*, the exploited? Tellingly, this was also a moment when racial resentment once again became politicized; the two most notorious EPP contracts were understood to link party elites and Asian businessmen.

EPP contracts leaching revenue at the headwaters of the system also made it difficult for Tanesco to enforce its monopoly on power consumers downstream. In 2010, a field survey estimated that Tanesco had suffered system losses of 24.4 percent (Azorom, AETS 2010). These combined "technical" losses occurred in the transmission and distribution process (12.8 percent) and commercial losses; that is, billing or meter irregularities (11.6 percent). From 2004 on, commercial losses have been hovering at around 12 percent, a doubling of their previous ten-year average of 5–6 percent (Vagliasindi and Besant-Jones 2013, 350). As I describe in chapter 3, Tanesco regularly dispatched "loss prevention" teams to inspect neighborhoods for metering or billing irregularities. They encountered various arrangements that diverted the upward flow of currency, including meter tampering, stoppages, slowdowns, bypasses, and surreptitious reconnection—often with the help of *vishoka* and their collaborators in Tanesco. Like the downward flow of current, the up-flow of payment was diverted and redistributed in ways that challenged underlying assumptions of reciprocity between utility and consumer, and ultimately citizen and state.

Challenged but—crucially—not vitiated. The collective circuit of the power network was partially redistributed at both the generation and consumption end, but at each point Tanzanians debated or articulated the limits of those redistributions. If it was one thing for a politician to take the proverbial 10 percent cut off an EPP power generation contract, it was another to take the whole thing. If it was one thing for households to rack up debt on their meters (even debt they might never pay back), it was another to bypass those meters. These distinctions preserved the core functionality of the power network, even as they mapped the ways that network accommodated itself to facets of Tanzania's changing social and political milieux. And in this way, they indicate how parasitism has come to be tacitly reconceptualized in Tanzania from something to be eradicated to something tolerated, within limits.

The road was blocked; I didn't see that you called; this text message was mistakenly delivered to me (but since it concerns you I figured I'd pass it along). People or things don't always arrive at their destinations, or arrive as intended. Like many truisms, this rides the line between the banal and the profound,[12] but it is the point of passage that links an older anthropology of the gift to science and technology studies, with its emphasis on associational networks, and to the "new materialism" with its emphasis on thingly recalcitrance to human designs (Bennett 2005; Latour 2005). One of the most charming and interesting theorists of this truism is Michel Serres, particularly in his work *The Parasite* (1982).[13] Drawing on cybernetic theories of channels and transmission, as well as European philosophy and folklore, Serres theorizes the parasite as any agent that acts as an interrupting third, perturbing a connection between two (or more) relata. His classic example is the fable of the country mouse guest who eats at the expense of his city cousin host, who in turn depends on the table scraps of a village tax collector, who in turn "takes a cut" of the productive efforts of his constituents, who themselves exploit the fecundity of the land, and so on down the line. In a cosmic sense, parasites are the generative principle of complex systems, branching out into an infinity of networks. "Life," as Eduardo Viveiros De Castro asserts, "is theft" (2014, 16).

In practice, parasites are always defined in relation to what media theorist Anna Watkins Fisher (2020, 25) calls the "threshold of accommodation." From Somali pirates along transnational shipping routes (Dua 2019) to the "corn jobbers" and "forestallers" that nestled into the highly regulated bread trade in early modern England (Thompson 1971), parasites change the "facts on the ground" by occupying some strategic position along a channel and nesting on its flow. But if the parasite exploits its host, for that very reason it must learn to play the good guest, keeping its presence within some tolerable margin; it cannot be too obtrusive or take too much. Conversely, the host must inevitably learn to accommodate the larger world of which that parasite is an emissary; it cannot get too "immunological," too preoccupied with purifying its environment (Esposito and Hanafi 2013). In short, guest and host must share the meal, such that some sort of ecological coexistence becomes possible.

This is easier said than done, of course. Within such coexistence, "the real parasite" is always a matter of perspective, even recrimination. Is it the guest who, uninvited, comes between the host and his meal? The host who, hoarding his stock, forces his guest to sing for her supper? Or is it the meal itself which, because it only goes so far, inevitably injects a sour note of material calculation into their amity? In these delicate circumstances, hospitality may turn to hostility (Shryock 2019). The balance of forces determining who gets what may well shift its equilibrium.

This is one way to understand the long modernizing century of increasingly networked flow. By the end of the nineteenth century, an emerging biopolitics in Europe and the United States sought to stabilize the runaway exploitation of urban industrial capitalism whose hard times were memorably captured with ethnographic realism by authors like Charles Dickens and Henry Mayhew. "Architecture or revolution," Le Corbusier ([1923] 2013) famously warned: civil society reformers and technocrats sought to make populations healthy and productive (and docile) through expanded provision of water, light, housing, and security. Meanwhile, labor wrested democratic power and legal protections, sometimes by exploiting critical chokepoints in the energy supply until their demands were met (Mitchell 2011). By *les trente glorieuses* (1945–1975) of postwar reconstruction and seemingly limitless energy consumption, the corporatist compromise between capital and labor in the white North Atlantic could generate a powerful sense of social security, if not boredom, albeit one paired with an ambient Cold War paranoia about infiltration and "vital systems" collapse (Collier and Lakoff 2015).[14] Here was a purified, modernist world in which money was mediated by impersonal bureaucracies and markets, in which life was regimented into vast formalized systems of banking, housing, and consumption, underwritten by a national state.

Decolonization in the social revolution mode marked Africans' utopian attempt to transcend their historical exploitation and assume this modernist form. Since the slave trade, the continent had been a source of raw materials that nourished modern comforts and marvels (D. M. Hughes 2017). It is not for nothing that Walter Rodney (1972, 179) calls the colonial epoch the "Age of Electricity," with African-mined copper lining the power grids strung across Europe and the United States. For educated young civil servants like Julius Nyerere that comprised the core of nationalist leadership in Tanganyika, this was not only a moment when the "straws of exploitation" could be cut and colonial parasites could be swept away, but one where modern standards of living might be available en masse and Africans might claim membership in a global humanity. Eager to cut history off at the pass, the centralized party-state nationalized the economy and attempted to route all flows of goods, money, and people through its institutions, adopting its own Cold War–inflected "security" discourse of eradicating parasites and enemies.

By the 1980s, however, it was clear that this utopian ambition belied an enduring postcolonial condition of underdevelopment and "unequal exchange" (Amin 1978). Most African countries remained "open economies" (Guyer 2004, 116–18), reliant on primary commodity exports for hard currency and unable to develop the industries required to spur economic growth and expand public infrastructures, which remained limited and patchy. Compounded by the austerities of structural adjustment, elites improvised forms of "private indirect government" (Mbembe 2001) that involved diverting funds flowing through state apparatuses and funneling them to intermeshed networks of security, extraction, and trade. Meanwhile, rapidly urbanizing populations "survived on the basis of markets that emerged spontaneously to recycle the money concentrated at the top and to meet the population's needs for food, shelter, clothing and transport" (Hart 2010, 375). In short, both accumulation and survival came to revolve around what Ferguson calls distributive labor: "processes of diversion, division, and tapping into flows" (2015, 96). And if everyone is to some extent or other tapping or diverting, then flow itself becomes a performative negotiation, a "conversion" across some sort of social or territorial differential (Guyer 2004). Different actors claim the right to levy rents on value circulation, a practice whose dual connotations of cutting (as in to rend) and connecting (as in to render) express its quintessentially parasitic logic. Rather than fading into the taken-for-granted background, the channels that make flow possible are reflexively figured, subject to contestation, reinterpretation, and multiplication of use. So long as they don't choke off circulation altogether, these negotiations over who gets "a piece of the action" serve as powerful sites of redistribution, tracing out the shifting contours of community, membership, and social obligation. Pushed too far, however, they can erode the basic sense of shared coexistence and fuel the kinds of civil conflicts that flourished in the 1990s.

Perhaps the most vivid examples of this dynamic involve the military checkpoints and roadblocks that have mushroomed across road networks in West Africa, Democratic Republic of Congo, and Central African Republic. From the perspective of drivers and travelers who are often subjected to security shakedowns and intimidations, these checkpoints are sites of extraction and predation. But as Janet Roitman shows in her ethnography of the Chad Basin (2005), for such *coupers de route*—a mix of militiamen, state officers, bandits, and other would-be regulatory authorities—working the roads is a kind of opening. It allows them to "pay themselves" when conventional salary flows from state institutions have been compressed or cut off. Roads are thus sites where basic questions about the nature of wealth and social obligation are broached: who owes what to whom? Who is the parasite and who is the channel? And what is the proper distribution of these intersecting flows of money and movement in conditions of scarcity and volatility?

The answers to these questions tend to emerge through experiment and improvisation. Consider a scene from McGovern's (2010, 188) "checkpoint ethnography" during a ramp up to potential civil conflict in early 2000s Cote D'Ivoire:

> On one particularly exhausting and hot minibus ride I was on, the security forces kept us a bit too long for the liking of several older women who were bringing their goods to a nearby market. They suddenly switched out of the mode of silent simmering anger we had all assumed, and began to berate the soldiers and police. "You are our sons. Why are you holding us up like this? There are no rebels in this bus you know that! We are your mothers and we are tired. We have been working since dawn, and we have places to go. Give us our papers and give our driver his papers before we curse you." After a slight pause that did not do much to help them save face, the soldiers decided that our papers were in order and we were on our way. Laughter and congratulations ensued in the vehicle as soon as we were out of earshot. Nevertheless, we all knew that such heroism succeeded in such a setting not because the women had justice on their side, but because so many other vehicles and passengers—indeed these same women at other checkpoints at other times—would suffer the delays and indignities visited upon them in silence.

While the tense situation of "neither war nor peace" in 2000s-era Cote D'Ivoire makes this example particularly charged, similar shakedowns can be found along roads across the continent (Takabvirwa 2018), including many I and countless others have experienced in Dar es Salaam by traffic police.[15] What it suggests is that while the parasitic redistribution of movement and money can unexpectedly explode into debates or even violence (Lombard 2013), there are also tacitly inferred (and occasionally explicit) principles that regulate its intensity, allowing it to remain within a tolerable "threshold of accommodation" that preserves the formal coherence of the relation. After all, not all parasites kill their host—only those who take too much. At sufficient levels of violence or expropriation—that is, when road-blockers are not simply uninvited guests to tolerate but enemies to avoid—passengers will reroute or stop traveling altogether.

Properly regulating parasitic redistributions thus entails making inferences and predictions. How aggressively should *coupers de route* comport themselves? How often can passengers press their luck and protest? How much time, money, or dignity would it save them? McGovern's phrasing—that market women "switch out of the mode" of silent endurance into something more confrontational (2010, 188)—is noteworthy; modality is a useful way to think about these calculations. The concept receives explicit treatment in philosopher Helen Verran's commentary on an influential work of Africanist anthropology: Jane Guyer's discussion of another performative negotiation, this time by a woman named Madame A, the proprietor of a petrol station in rural Nigeria. Madame A was able to honor the government-mandated price of petrol during a supply shortage while tacitly allowing the "real price" to rise through ancillary discounts and premiums (time waited, anxiety, volume of product), thereby preserving a certain "range of normality" (2004, 114). As Verran (2007, 181) describes, Madame A's virtuoso performance both exemplifies and invites a "modal reasoning" that grasps the integrity of the whole amid the shifting arrangement of its parts:

> We might find ourselves saying: "Generally, the price of petrol goes up in a shortage." The adverbial way we use *general* here can alert us that we are engaging modal reasoning. This form of generalization evokes a vague whole that we might call the "petrol economy." The specific ordering of some parts and the general order they constitute are articulable. Relational empiricism studies the "lives" of these vague wholes, specifying the ways its parts come to life and perhaps die off, identifying the mediations that are important in the "doing" of this vague whole.

In both linguistics and philosophy, modals are operators broadly concerned with grammatically or logically qualifying a statement by expressing its counterfactual or nonactual dimensions—with the interplay of possibility and necessity (epistemic or "natural" modality), permission and obligation (deontic or "social" modality). In his own discussion of modal auxiliary verbs, Kockelman gives the following example: "while you *may* use a teaspoon to dig a ditch (insofar as no will arrest you) you probably couldn't dig that ditch very fast or very steep" (2017, 131). Put another way, modals express an inferred range of possibilities, and their various tradeoffs, of an emergent relation (say, the channeling of dirt) in general. A spoon is one mode of digging, hardly ideal but one that still qualifies. Other modes are not discrete but continuous (and to capture this fractional quality we might use the more diminutive term *modi-*

fication): a spoon with a bent handle would also count as a mode of digging, one that increasingly (dis)qualifies, but does not vitiate, its basic character. Finally, while modals are inherently attuned to other possibilities and hence have a disruptive quality (they gesture subjunctively to "being otherwise"), they also have an important link to *modesty*. They always proceed *in medias res*, reflecting on what one could or should do with the materials at hand—on how actor and environment mutually impinge upon each other to form a distributed agent with its own logic and particular range of movement.[16]

Since the economic crises of the 1980s, Dar es Salaam has been popularly known as Bongo, a Kiswahili word that literally means brains, and more figuratively refers to the way one must use one's ingenuity (*ujanja*) to survive or prosper (Callaci 2017). In this way, it is part of a larger cultural vocabulary of cleverness and cunning that has proliferated across African cities in the wake of structural adjustment (Petit and Mutambwa 2005). In showing how politicians, residents, electricians, and utility workers all work the channels of current and currency flow, I build on this rich literature of African urban life "after adjustment" (Melly 2017; Simone 2004a; De Boeck and Plissart 2004). But by thinking about these "parasitic" exploits in terms of modes and modification, I also want to highlight the underappreciated role of *modesty* in urban life, by which I mean the ways these exploits heed a larger collective relation and thereby, at least for the moment, remain within its form.

To flesh out this idea, we can turn to Gregory Bateson's own well-known example of moving in concert with one's environment: "Consider a man felling a tree with an axe. Each stroke of the axe is modified or corrected, according to the shape of the cut face of the tree left by the previous stroke. This self-corrective (i.e., mental) process is brought about by a total system, tree-eyes-brain-muscles-axe-stroke-tree; and it is this total system that has the characteristics of immanent mind." (1972, 230)

Cutting here is continuously modified, shifting force and angle, but in a regulated way whose range of variations corresponds to the notches in the tree. The woodcutter cannot hack blindly at the problem, but must align his eyes and muscles to the wood, to see it from the wood's perspective, as it were. In turn he enters a form whose logic (or mind) is distributed across its component parts. Likewise, for those residents, politicians, or "hatchets" looking to successfully modify the circuit of current and currency, either to exploit or repair its form, a similar sort of modesty is required. It requires not being too greedy or thoughtless, of heeding the basic premises of collective interdependence that, like road networks, the power network manifests. As I show across the following chapters, it is not quite right to say that actors who parasitically divert resources away from collective circulation are subject to moral re-crimination. Given the centrifugal political economy of liberalization, this is only to be expected. The problem occurs when such actors *take too much*, when they foolishly or greedily ignore various subtle distinctions between what is tolerable and what is insensible—that is, between rent and plunder, load shedding and blackouts, debt and theft. By contrast, to reason modally, to model the consequences, is to think with and as part of the system itself as it unfolds into the future.

The chapters that follow explore postsocialist modes of generation, transmission, consumption, and maintenance/extension. Each is a "quali-fied" expression of a given stretch of the infrastructural circuit. Each mode accommodates some parasitic interference, whether in the downward flow of current (oligarchic rent-seeking at the generation level, supply deficits at the transmission level), or in the upward flow of currency (Tanesco's difficulties in enforcing payment in the labyrinths of popular neighborhoods, the insertion of unofficial *vishoka* brokers into Tanesco's bureaucracy). Put another way, they all express the network's inextricable enmeshment with other networks that constitute its environment: Tanzania's politics and publics, its landscapes and livelihoods. And they express how, as the power network came to be the symbolic and practical center of Tanzania's national compact, actors had to strike a balance between undermining and preserving it.

the chapters

Consumers' access to electric current depends on the household wiring it flows through, which in turn depends on the public service lines that feed it; all of which in turn depend on the branch and root Tanesco offices—the institutional bureaucracy that regulates the physical network as such. To borrow anthropologist Nancy Munn's description of the Kula Ring (1976), this is a nested hierarchy of spatiotemporal expansion, and the different scales demand different methods. As I show in chapters 1 and 2, the headwaters of the network require an ethnographic analysis attuned to elite political economy, as well as the shifting ideological and affective currents of public life as a whole. Chapters 3 and 4 fan out to the decentralized "tail" of the network at street level. Here the object is less the overt citizen/state relation than the utility/consumer relation that forms its derivative, and concerns the landscapes and livelihoods of urban neighborhoods. While the chapters move in one direction only, each level is a bridge up to the one it succeeds.

Chapter 1 ("Emergency Power: A Brief History of the Tanzanian Energy Sector") begins at the headwaters of the network to examine the way large scale changes in Tanzania's political and economic systems intersected with the generation of electricity. A robust Tanzanian power sector is a key prerequisite for economic growth and stability, but the upstream processes of allocating funds and tendering contracts has often been parasitically diverted to other ends. Focusing on these upstream conditions not only explains much of the expense and shortage in which Dar es Salaam electricity consumers currently find themselves, but highlights what we might call actually existing privatization and the forms of post–Washington Consensus oligarchy it is bound up in.

Since the 1990s, cyclical droughts have prompted dubious government tenders to well-connected private companies for emergency infusions of oil-generated electricity. These public bailouts are quickly converted to private rents that in turn feed the patronage network and fund electoral campaigns. The chapter focuses on two key instances of this dynamic: the 1996 contract with the Malaysian-Tanzanian company Independent Power Tanzania Ltd, for 100 MW of heavy fuel oil generated electricity, and the 2006 contract for 120 MW of thermal power with Richmond Development, an ostensibly American company with direct ties to the prime minister at the time, Edward Lowassa. Both of these supposedly short-term emergency generation arrangements were incorporated into the grid's long-term functioning, and both helped to cripple Tanesco's operations.

Despite or because of the damage these contracts inflicted, a residual sense of political centralization remained. In response, the sitting presidents Benjamin Mkapa (1995–2005) and Jakaya Kikwete (2005–2015) facilitated a series of anticorruption sweeps to hold the worst excesses of their party's elites in check. Still, rather than seamlessly regulating the tradeoffs between acquiring private rents and retaining public legitimacy, the rhythm was that of a yoyoing *fort/da,* of punctuated power cries and annual rationing periods, offset by occasional anticorruption sweeps and new jolts of expensive thermal generation. I have come to think of the interweaving of politics and power, elections and electricity, as a single system of emergency power in which the ruling party CCM incrementally burned through on-demand oil contracts, smash-and-grab rents, and residual socialist legitimacy.

Chapter 2 ("The Flickering Torch: Power and Loss after Socialism") follows electricity from its generation sources as it is distributed (or not) throughout the city. One of the most damaging consequences of the emergency power contracts and sector reform more generally is that it proved unable to overcome the grid's generation deficit, locking in a cycle of sporadic power interruption capped off by periods of intense, countrywide power rationing. The bulk of 2011 was one of the worst such periods and coincided with the beginning of my long-term fieldwork.

Drawing on scholarly work attuned to the poetics and semiotics of infrastructure, this chapter describes how Dar es Salaam's public read and responded to these different sorts of power loss. When properly—that is to say publicly—distributed, power loss bound the city together in a shared atmosphere of sacrifice that, I argue, reprised core elements of Tanzania's political culture. Since *ujamaa*, Tanzanians have regularly been called into existence through lack, asked to suffer collectively in exchange for the promise of a future fullness. Key here was the ability of words to act as a placeholder for things; media and public communication more generally in Tanzania have long functioned as a kind of second-order phatic infrastructure that enacts the form of a social relation when the vital flows that substantiate it are interrupted or delayed. Indeed, all manner of ideological broadcasts accompanied the power crisis of 2011, drawing contextualizing links between past and future, cause and effect, and generally re-inscribing it within a developmentalist narrative. That said, promises and explanations only go so far. After fifty years of independence, amidst a springtime of popular uprisings, yet another call to endure national hardship by an incumbent party grown rich could seem but a faint or distorted echo of more optimistic times.

Such skepticism, moreover, was only reinforced by the kinds of power cuts that could *not* be plausibly narrativized. From 2012 onwards, irregular or unexplained cuts frazzled the public, giving rise to rumors and suspicions about covert and illegitimate rationing, and resonating with a wider "communication breakdown" marked by the forceful silencing of political opposition. Enduring these shifting patterns of power outages and their effects on the public nervous system, residents articulated an important and key postsocialist distinction: if it is one thing to endure absence, it is another to endure it in the absence of explanation.

Chapter 3 ("Of Meters and Modals: Patrolling the Grid") descends down to the street level of urban life to examine electricity's point of sale and consumption. For resource delivery infrastructures such as power (or water), the meter marks the switch over from producer to consumer, and from current to currency. However, this switch is not a straightforward process. The technical structure by which meters record use, the larger bureaucratic procedures to which they are attached, and their enmeshment in urban space all create potentials for diversion. On the utility side, current arrives more expensively and less reliably thanks to the forms of emergency power that generate it upstream, and this stokes consumer dissatisfaction and distrust. On the other hand, payment itself must be consolidated and rendered upwards by households that are themselves plugged into an informal economy and landscape that disincentivize even rhythms of payment.

The chapter offers an ethnography of Tanesco disconnection teams and revenue protection units as they circulate through various neighborhoods to monitor, inspect, and occasionally disconnect household and commercial meters. In narrating these patrols, the chapter describes the formation of two recalcitrant landscapes of electricity use in postsocialist Dar es Salaam: an older, denser, and poorer urban core, comprising what are known as "Swahili" (*uswahilini*) neighborhoods, characterized by multiple-family rental housing and an aging postpaid metering system; and a relatively new, spacious, and wealthy set of neighborhoods characterized by gated compounds with prepaid meters. Residents in both diverge from Tanesco's ideal of legible citizen-consumers who pay for service in even and transactionally precise cycles. Instead, high levels of debt and surreptitious reconnection tend to characterize the former neighborhoods, while incentives for intentional theft and tampering tend to characterize the latter. If the official socialist teleology would have the poor Swahili classes eventually join the white-collar clerks "on the grid" as responsible citizens, the enterprising indiscipline of wealthier compounds suggests something else has unfolded. In certain ways the latter have drifted to more "Swahili" styles of life that anticipate scarcity and strategically disengage from projects of state legibility.

Faced with the evasions, protests, and obstructions of those who do not wish to be disconnected for debt or theft, some inspectors rail against customers who want it "easy" with stolen power or unpaid bills, echoing a socialist discourse of discipline and hard work. However, patrols are also well aware that the same liberalizing forces that created this indiscipline press upon them as well, in the form of diminishing pay, equipment, and job security. Some inspectors incorporate extortionist or protectionist arrangements with customers, while others maintain an ethical outlook steeped in the "socially thick" Fordist labor regime that Tanesco could still resemble even in the 1980s and 1990s (Ferguson 2006, 197). Somewhere between rejecting and exploiting the putatively "Swahili" mentality of easy money, Tanesco patrol teams and customers collaboratively exercised a kind of modal reasoning about what kinds of diversions of payment are tolerable and which ones are insensible.

Chapter 4 ("Becoming Infrastructure: *Vishoka* and Self-Realization"), finally, turns to struggles over maintenance and extension. After years of dwindling investment or funds for customer service, access to a functioning service line is beset with all manner of difficulties. *Vishoka* like Simon have emerged to facilitate access, expedite customer applications, provide emergency repairs, tamper with meters, or divert materials and supplies to residents in parallel markets, often by collaborating with Tanesco employees "inside" (*ndani*) the institution.

Drawing on a rich Africanist literature on wealth in people, this chapter shows how *vishoka* livelihoods and careers are reincorporated into Tanesco's customer service processes. In general, *vishoka* are incentivized to move upstream from the relatively "technical" decentralized edges of the network inward to its relatively "social" center, from small-scale repairs and reconnections at the street level to the second-order sphere of mediating the bureaucratic process by which the residents connect to the grid. This arc requires them to build up singular reputations as trustworthy collaborators to both consumers and Tanesco bureaucrats alike, which is not always an easy task. In this way we could speak of their careers as a process of "becoming infrastructure," of inserting themselves as mediators of the network at increasingly deeper, more "social" levels. Both parasite and channel, they are the densest expression of Tanzania's postsocialist condition as a living circuit, a give and take of mutual adjustment and responsiveness that threatens to fall out of form; but, at least in the first decades of the twenty-first century, managed to keep spinning.

Solving the power crisis is rapidly becoming the defining problem of the fourth phase administration, in much the same manner as nationalizations and Ujamaa villages defined the first phase, liquidations of parastatals the second, and failed privatizations the third phase.   —NIMI WETA

emergency power   1

A BRIEF HISTORY OF THE TANZANIAN ENERGY SECTOR

With burnished bleachers and long vertical windows that glow with sunlight, Karimjee Hall has a stately aesthetic well suited to deliberation. In 1955, the Hall was built and donated to the colonial government by the philanthropic Karimjee family, whom a recent historical retrospective dubs "the Merchant Princes of East Africa" (Oonk 2009). At Independence in 1961, it was the setting for Nyerere's inauguration ceremony, and housed Tanzania's National Assembly until the 1970s, when Nyerere moved the capital to the geographical center of the country, Dodoma, symbolically rejecting the commercial coast for the socialist heartland. But while Parliament convenes in Dodoma, most of the ministries remain in Dar es Salaam, and over the years Karimjee has hosted a grab-bag of state, donor, and NGO functions (e.g., Mercer 2003, 757). Walking into the space one early December morning during my fieldwork in 2012, I could sense this accumulated history in its mishmash of infrastructural scarring. An enormous digital projector screen glowed blue behind the elegant wooden dais, and a row of boxy AC units jutted out incongruously from the windows. On that morning those units were on full blast in anticipation of the day's event. The Energy and Water Regulatory Authority (EWURA) was holding a town hall meeting to discuss the price of electricity.

New and well funded (established in 2006), EWURA faithfully held these sorts of "stakeholder consultations." They are part of the suite of "best practices" thought key to encouraging civil society. But, as the master of ceremonies noted with disappointment that morning, the event was sparsely attended, despite the radio and newspaper promotion. His audience—a handful of representatives from Tanesco, the Ministry of Energy and Minerals (MEM), and ZECO (the Zanzibar Electric Company) speckling the seats—ruefully chuckled. Their attendance was good P.R., at least. Two enterprising photographers crept about the bleachers with digital cameras, snapping pictures of the dignitaries. Later, I would walk out into the lobby to find glossy 3 × 5 printouts laid out on the floor, conveniently available for purchase.

The sparse attendance would not mean much in the end. The meeting was a minor episode of a larger, more protracted conflict. A year before, in 2011, a drought had prompted Tanesco to submit an emergency request to increase electricity tariffs by 155 percent, from Tsh. 195 to Tsh. 497 for domestic consumers per unit; EWURA granted a provisional 40 percent increase, withholding final approval until Tanesco undertook a cost of service study that could justify the request.[1] The Ministry of Energy and Minerals hired the Spanish consulting firm Mercados, which recommended a succession of moderate increases over three years: Tsh. 263.65 in 2013, Tsh. 268.03 in 2014, and Tsh. 291 in 2015. These tariffs would be additionally subject to automatic adjustments to take into account fluctuations in currency, inflation, and fuel prices. Nevertheless, these hikes were substantially less than Tanesco had claimed it needed. Now, a year later, EWURA was ready to review the recommendations, and requested the input of all relevant "stakeholders," including the public, such as it was.

After introductions, the morning began with the Tanesco acting manager pleading his case. The increases did not go far enough, principally because the consulting firm's calculations were predicated on a number of faulty hydrological assumptions. After years of poor climatic conditions, the water levels in the nation's hydropower network were depleted. The level at one of the main dams, Mtera, was 687.78 meters above sea level, just below the minimum threshold needed for generation, 690 meters. To counter this deficit, Tanesco was relying on the purchase of expensive thermal generation from emergency power producers (EPPs). Unless consumer power tariffs rose dramatically, Tanesco's debt of $250 million to these suppliers would shoot up to $800 million.

As self-evidently pressing as these numbers were, Tanesco did not have reason to be optimistic. Earlier that week the head of the MEM had blithely announced to the media that there would be no price increases. Tanesco's fiscal problems, he suggested, lay in the utility's own financial mismanagement, and the consumer should not be punished for that. "The consumer" was in full agreement; as the morning progressed, members of local media and the ever-elusive civil society did trickle in, taking seats near the back entrance of the hall. After Tanesco, MEM and EWURA gave their presentations, the floor was open to questions and comments, most of which expressed disapproval of the hiked tariffs. A middle-aged man stood up:

> Salaam aleikum. I don't have much of value to contribute to this debate, except what I will discuss. My name is Othamn Ramahdahni and I'd like to discuss the weakness of Tanesco. Tanesco has come here to raise their bills [their price] when they themselves cannot provide service, but lust after the money of Tanzanians. Tanesco, the things they do . . . for example, I live in Temeke. The point will come when you request LUKU [a prepaid meter]. I've been surprised to hear the Tanesco manager here today say that Tanesco is now able to install LUKU meters when it's not true. If you come to [the branch office] Temeke, everyday you're told the LUKU meters have not arrived, the materials aren't there, and whatever else. You have the documents, you've already paid, but power in your house? There is none.

Besides a few appreciative chuckles, Othman's comments were met with awkward silence. His agitation was patently at odds with the measured tones that had otherwise filled the hall, as was his discussion of meters instead of dams. Like most consumers, he perceived Tanesco from the distribution end, as a series of cryptic blackouts and rising prices. In what world could this justify the steady price increases of the past decade? Indeed, two months later, the MEM "advised" that Tanesco should withdraw its application and maintain the 40 percent increase.

Minor though this episode was, it is instructive for understanding the multiple logics that characterize the actual workings of utility governance. Despite the commissioned, in-depth technocratic study, the determination of Tanesco's tariff was, in the final instance, paradoxically both pragmatic and fantastical. On the one hand, the MEM had strong-armed Tanesco into withdrawing a politically unpopular increase. On the other hand, it was kicking down the road Tanesco's inevitable fiscal reckoning. This pragmatic negotiation over the increase of currency flowing "upwards" to Tanesco, relative to the current it was supplying, I will argue, leads us to the heart of the trends in Tanzanian statecraft over the last two decades. It indexes the ways the realignments of elite politics have shaped the workings of the national power sector, and the way the power grid has become a site to interrogate and resist those changes. By 2011, entrepreneurial elites in the ruling party had learned that they could—and perhaps even had to—parasitically extract value from national patrimony like Tanesco. But they also learned to modulate that extraction so as to avoid excessive political fallout. In this vein, electricity can provide a rich medium for an ethnographic portrait of an African state in the contemporary era, whose broad yet unifying spirit Charles Piot (2010) locates "after the Cold War," and that in Tanzania's case specifically, we may designate as postsocialist.

In this chapter, I describe the stretch of Tanzania's postsocialist era spanning 1995–2015 as one of emergency power. The term literally refers to the aforementioned services of emergency power producers (EPPs), private companies temporarily contracted to import liquid-fuel generated power to supplement Tanesco operations when its hydropower sources are dangerously low. But I will argue that EPPs indexed "emergency power" in a more expansive sense, a point of entry into a whole interlinked condition that involved the specter of Tanesco's fiscal meltdown, the physical energies of thermal generation, the political energies of a ruling class adapting to the pressures of intra- and inter-party competition, and in turn, the cultural mood of Tanzanian postsocialism. The remainder of this section elaborates this point, while the following introduces two companion terms—the air conditioner and the veranda—as a way to contextualize emergency power within long-running, though as yet unresolved, debates about the nature of neoliberalism. The middle third of the chapter tracks the evolution of Tanzania's power sector through the colonial era, the high modernist socialist periods of the 1960s and 1970s, and the crisis of the 1980s. The final third drills down to the details of power sector reform in the late 1990s and 2000s, its (racialized) politics and scandals, its dynamics of legitimacy and accumulation.

What then is emergency power? At first glance, emergency power *producers* are a category of electricity generation, part of a larger mix that includes Tanesco-owned plants (mostly hydropower) and independent power producers (IPPs)—companies with long-term contracts to generate and sell electricity, and who, it has been expected, could one day play a permanent role in a fully privatized power sector, selling electricity to private wholesale distribution companies. The IPPs that came online during the 2000s also tend to use natural gas, and thus occupy a middle ground between state-owned hydro-dams and rented emergency liquid-fuel generators (usually diesel or heavy fuel oil). If we look at statistical representations of the fuel mix, the influence of EPPs would seem to be marginal (see Table 1.1). By May 2015, Tanesco's on-grid installed capacity was 1,525 MW.[2] Of this capacity, Tanesco-owned sources comprised the majority with 883.2 MW (~58 percent). This was closely followed by IPPs and small power producers (SPPs) who together comprised 437 MW (~28 percent).[3] Emergency power producers accounted for a mere 205 MW (~13 percent). Moreover, after 2015, EPP contracts (Symbion and Aggreko) were not renewed, partly due to improved hydrological conditions and partly due to the construction of new, state-owned Kinyerezi gas plants I (150 MW) and II (240 MW) in late 2015 and 2018, respectively.

During the 2000s, when donors, consultants, and other development in-dustry actors tried to turn this table into a narrative, they often came up with a baby and bathwater story: the government of Tanzania should throw out the residual EPP bathwater while keeping, and growing, the healthy IPP baby.[4] For most of its existence, the Tanzanian power sector was a vertically integrated monopoly composed of generation, transmission, and distribution components, each owned by the state power utility Tanesco. Starting in the 1990s, Tanesco signed onto the broad process of neoliberal power reform along market principles, beginning with contracting IPPs to sell privately generated power that it would then transmit and distribute to consumers. For the most optimistic of reformers, this would culminate in private investors purchasing Tanesco's government- and self-owned shares, as well as selling off the distribution network and generation assets such as the Mtera Dam (Kasoga 1998). In the meantime, Tanesco itself would "prepare" for full priva-tization by instituting a number of commercial reforms to get in the black and appeal to investors, including instituting cost-reflective tariffs and enhancing operational efficiency.

Unfortunately, the story goes, a combination of bad luck and bad actors knocked this process off course: at critical junctures (1992, 2005, 2011), water levels in national dams became dangerously low, necessitating the use of EPPs, which should have been a temporary, stop-gap solution. Instead, government officials awarded lucrative and often long-term contracts to EPPs, one of which was little more than a fictional "briefcase" company. As a result of this cor-ruption, Tanesco's financial and technical underperformance compounded as these noncompetitive tenders absorbed revenue that otherwise might have been reinvested into improving the distribution network or payment and billing services. In short, EPPs were parasitic. They siphoned value along a crucial chokepoint of the energy supply chain, derailing the legal-rational privatization process and delivering consumers the worst of both the socialist

TABLE 1.1 On-Grid Generation Capacity, ca. 2015

Plant Name	Fuel Source	Capacity	Ownership	Commission Year	Retirement Year
Hale	Hydro	21MW	TANESCO	1965	2038
Nyumba ya Mungu	Hydro	8MW	TANESCO	1967	
Kidatu	Hydro	204MW	TANESCO	1975	2025
Zuzu	Diesel	7.4MW	TANESCO	1980	2014
Mtera	Hydro	80MW	TANESCO	1980	2038
Uwemba	Hydro	0.8MW	TANESCO	1989	
Pangani Falls	Hydro	68MW	TANESCO	1995	2045
Tanwatt	Biomass	2MW	SPP	1995	2029
Kihansi	Hydro	180MW	TANESCO	2000	2050
Tegeta IPTL	Heavy Fuel Oil (HFO)	103MW	IPP [originally EPP]	2002	2021
Songas 1	NG	38MW	IPP	2004	2023
Songas 2	NG	110MW	IPP	2005	2024
Songas 3	NG	37MW	IPP	2006	2025
Tegeta NG	NG	44MW	TANESCO	2009	2028
Tanganyika Planting Corporation (TPC)	Biomass	17MW	SPP	2010	2030
Ubungo I	NG	102MW	TANESCO	2010	2026
Symbion Ubungo [formerly Richmond/ Dowan]	NG/Jet Fuel	126MW	IPP [originally EPP]	2011	2014
Aggreko Tegeta	Diesel	50MW	EPP	2011	2015
Aggreko Ubungo	Diesel	50MW	EPP	2011	2015
Ubungo II	NG	105MW	TANESCO	2012	2031

Plant Name	Fuel Source	Capacity	Ownership	Commission Year	Retirement Year
Mwenga	Hydro	4MW	SPP	2012	2062
Symbion Arusha	NG/Diesel	50MW	EPP	2012	2015
Symbion Dodoma	Diesel	55MW	EPP	2012	2015
Mwanza	HFO	63MW	TANESCO	2013	2038
Total as of 2015		1525.2			

SOURCE: Statistics compiled by author from Mkobya et al. (2013); Lennart et al. (2014); International Law and Policy Institute (2014); Eberhard, Gratwick, and Kariuki (2018).

and capitalist worlds: the odious mix of high prices *and* inefficiency that Othman Ramadhani denounced.

The story of a corrupt African state disappointing a compliant Western corporate/donor class is a familiar one, but as Hannah Appel (2019) has shown so well in her ethnography of a US oil company in Equatorial Guinea, such distinctions emerge through deeper complicities and entanglements. If we look closer, we see that some African companies like International Power Tanzania Limited (IPTL) started out with EPP contracts and morphed into IPPs, while American companies like Symbion opportunistically expanded into temporary EPP operations. If we bracket distinctions between "normal" IPPs and "pathological" EPPs, we simply see the increased privatization and thermalization of power sources in general. And this framing suggests that any such differences must themselves be understood as effects of a broader field of "neoliberal" transformation.

the air conditioner and the veranda

The beginnings of Tanesco reform in the 1990s coincided with a market triumphalism that seemed to herald a global "privatization of everything" (Watts 1994). This, some imagined, would invigorate liberal-democratic civil society (Huntington 1993), transition the former Soviet bloc away from socialism, and bring the world into an "End of History" steady state (Fukuyama 1989). Three decades later, the hubris is easy to see. Instead of the bell curves of mid-century modernity, what proliferates is J-shaped, "power law" distributions (Hart 2004; Guyer 2010). Global institutions are run by "flexians" who move in and out of government and private sectors with ease (Wedel 2009), while increasingly precarious populations are subject to structural unemployment, self-responsibilization, and incarceration (Han 2012).

While there are many overlapping ways to sort the literature on these neoliberal developments (Gledhill 2018; Hilger 2011), one useful axis spans relatively Marxist versus relatively Foucauldian approaches (Wacquant 2012). Marxists emphasize neoliberalism as a political project of class restoration via liberated restrictions on the movement of capital (Harvey 2005). Foucauldians, by contrast, see a mobile and mutable "rationality" articulated by intellectuals, policymakers, and other experts (Rose 1996; Ong 2006; Collier 2011). The former has been critiqued for a hermeneutics of suspicion that sees any and all phenomena as emanations of a totalizing, Leviathan-like condition (Collier 2012); the latter for a polymorphous, assemblage-style particularism that can only traffic in "an indefinite number of small-n neoliberalisms born of the ongoing hybridisation of neoliberal practices and ideas with local conditions and forms" (Wacquant 2012, 70).

The view from Africa suggests we might employ these two perspectives historically (Hilgers 2012). The first wave of 1980s-era neoliberalism involved the macroeconomic shock treatments of structural adjustment, austerity, privatization, and deregulation. The result was often military dictatorship and what Emmanuel Terray (1986) calls a "politics of the veranda," characterized by patrimonial networks indifferent to the rule of law. It is not difficult to see how this "criminalization of the state" (Bayart, Ellis, and Hibou 2001) aligns with a broader capitalist trend of class restoration after the social-democratic compromises and postcolonial developmentalism of the mid-twentieth century (Harvey 2005). And the "restoration" element should not be underemphasized. The first wave of neoliberal reform "was, in fact, largely a matter of old-style laissez-faire liberalism in the service of imperial capital . . . a crude battering open of third world markets" (Ferguson 2010, 173). With its oligarchic-corporate elite (Kapferer 2005) and multipolar imperial jostling, Africa after the Cold War harkens back to the predatory "antimarkets" (Braudel 1992; Wallerstein 1991) of warlords and robber barons in the late nineteenth century, the Gilded Age in which the "merchant princes" of the Karimjee family first made their fortunes. The offshore oil concession, Ferguson (2005b) suggests, is not so different from the Belgian Free State.

By the mid-1990s, however, this rightward drift epitomized by Reagan and Thatcher in the North Atlantic was tacking center to a softer Clinton/Blairite neoliberal "rationality." Against the backdrop of failed structural adjustments, flourishing tax evasion and shadow economies, and the rising tide of democratic aspirations and rights culture, international development discourse turned its focus to technocratic issues of good governance and state-*building*. An explosion of NGOs and philanthropic organizations took on the tasks of social policy in so-called developing countries (Rafiq et al. 2022), while the World Bank and other donor-sponsored reformers emphasized the need for regulated and competitive markets (or market-like mechanisms) to engender calculative rationality that would optimize sluggish bureaucracies, neutralize regulatory capture, and reduce the transaction costs of business and governance.

Call it a "politics of the air-conditioner" (Terray 1986; Kelsall 2002). Like many reformisms, it is often anemic, narrowly focused on procedure and technocratic fixes. Sometimes these techno-fixes are obviously (anti)political, as when prepaid meters in South Africa reinforced apartheid legacies of black segregation and marginalization (von Schnitzler 2013). Other times they are just ineffectual.[5] Still, as Collier (2012) has argued, at its best this version of neoliberalism emphasizes modest interventions, reflexive adjustment, and a detailed empirical engagement with the rigidities and frustrations of bureaucratized social states (see also Redfield 2016). A rich body of ethnographic work on the contemporary African state has shown how thinned-out social welfare measures such direct cash payments (Ferguson 2015), privatized customs regimes or entrepreneurial toilet complexes (Chalfin 2010, 2017), and subcontracted construction projects (Mains 2019) have enhanced the commitments to functional public services insofar as they deliver a measure of reliability to daily life, and thus legitimacy to the state.

If neoliberalism means anything in postcolonial Africa, it must refer to the layering of air-conditioner technopolitics over the veranda-style patrimonialism wrought by structural adjustment, and the web of tensions and complicities that emerges. This was certainly the case for Tanzanian power sector reform. While the shadowy dealings of certain EPPs contravened donor visions of a formalized, air-conditioner commercialization/privatization process, they were nevertheless consonant with broader shifts toward a patrimonial world of decentralized rent-seeking unleashed by structural adjustment a decade before. At least two major EPP/IPP contracts—IPTL and Richmond/Dowans (later Symbion)—provided a kind of "panic rent-seeking" (Cooksey and Kelsall 2011) for powerful oligarchs who had grown rich in the liberalization era and used the money for their own personal patronage networks and to finance costly reelection campaigns.

As we shall see, the strange stalemate between air-conditioner and veranda politics owed, in part, to the ways each could draw on different parts of the socialist legacy. As Othman Ramadhani's comments imply, technocratic fixes like prepaid LUKU meters in fact proved quite popular, since they extricated households from dependence on meter readers who *should* be serving the *umma*, the public good, but who had grown inefficient and predatory. At the generation level, emergency power scandals married popular resentment against capitalist exploitation to the better instincts of a neoliberal audit morality. They generated anticorruption sweeps, legal arbitration, and media denunciations of a ruling party that had seemingly abandoned its constituency for the financial support of a wealthy (if often tacitly read as "Asian") business class. There were, in other words, genuine resonances between an air-conditioner politics of good governance and a socialist commitment to moral rectitude.

At the same time, those self-enriching elites "out on the veranda" who in effect sabotaged the nation's power supply could also strike a socialist pose by wrapping themselves in the mantle of national sovereignty and postcolonial "self-reliance." Controversial EPPs were defended as "South-South cooperation" between African and Asian business leaders against the patronizing meddling of the World Bank and their agents. This posture of national sovereignty culminated in the 2005 election of Jakaya Kikwete and the formal rejection of the World Bank's privatization plans for Tanesco and much else besides. The discourse was one of resurgent nationalism, and varieties of private involvement with energy, water, and transport parastatals were all shut down. In preventing "foreign" control of the economy, these policies harkened back to Nyerere's socialist Arusha Declaration in 1967 and his resistance to structural adjustment in the early 1980s. They were meant to signal that CCM, despite its enthusiastic embrace of air-conditioner reforms during the Mkapa administration (1995–2005), could still have the interests of the nation at heart.

It is important to take a nonreductive approach to this anti-privatization mood, which had many legitimate practical and ideological justifications swirling about it. From 2002 to 2006, Tanesco was temporarily run by the World Bank-funded, private South African company NetGroup Solutions. NetGroup, with its brutal tariff increases and lucrative bonus contracts, often fell short of the better neoliberal traditions of combating regulatory capture, as we will see. Nevertheless, it is also important to situate the rhetoric of anti-privatization within a broader historical context in which increasingly bold rent-seeking among CCM elites was strategically vital but politically costly. Indeed, while most official privatization of parastatals was "patriotically" delayed or rejected by 2005/2006, CCM officials and financiers were still forging a de facto privatization of the state through all manner of no-bid contracts and "tenderpreneurship." The irony was not lost on commentators, including an aged Nyerere, who reportedly observed of the IPTL deal: "If this is an example of South-South cooperation, then colonialism was better" (quoted in Kabwe 2014).

From 1995 to 2015, then, the power sector was characterized by a series of veranda-style deals checked by legal redress and anticorruption sweeps; and conversely, by a series of air-conditioner reforms undercut by halted privatizations and nationalist posturing. This looping, figure 8 pattern kept the sector suspended in what one manager described to me as a mode of "reactive, crisis-driven planning." And yet there are good reasons for why and how this spinning in place came to be. It can be plausibly understood as an instance of what political scientists call "dysfunctional institutional equilibrium" (Godhino and Eberhard 2018, 33). While I would take issue with the patronizing and overly normative tones of the term "dysfunctional," especially in relation to an African postcolony, the concept in essence does describe something quite familiar to anthropologists: the way that failure itself can generate useful or consequential effects that benefit actors.[6] Tanesco may be "dysfunctional" insofar as it is not profitable in a narrow economic sense, but it nevertheless anchors a "political settlement" (Khan 2018) that has allowed Tanzania's socialist party-state to adapt the imperatives of liberalization. As we will see, the cumulative and often repetitive cycles of power reform mean that "at a system level there is an unexpected stability—one that has been able to keep CMM in power since the 1960s, perpetuates a deep political culture tied to the ideology of African socialism, and where the tensions between informal and formal institutions creates a sort of balance that serves the interests of a . . . political economic elite" (Godinho and Eberhard 2018, 33).

A key concern of this book is to explore how striking this sort of balance involves a modal reasoning about the appropriate limits of entrepreneurial liberty in a (post)socialist world. As journalist and researcher Brian Cooksey observed in his authoritative account of IPTL, "it is one thing for politicians and bureaucrats to take a cut from a valid investment that generates significant employment, turns out useful products, and contributes to government revenue. It is quite another for this group to take a corrupt cut from a project that derails a key national policy, and imposes huge additional costs on end-users and tax payers" (2002, 73). The very concept of rent, after all, entails both cutting *off* (as in rend) and connecting *to* (as in render). Rent may render access to a service, particularly in a decentralized institutional and political environment. Those rents might even loop back into the overall economy of state-society relations through other points of reentry and conversion. Pushed too far, however, they may simply rend the moral pathways linking rulers to ruled.

Electricity has been a particularly generative medium for experimenting with this dynamic equilibrium. While there was a run of "grand corruption" scandals in the late 1990s and early 2000s (Hazel 2015), I would suggest that those in the power sector deserve to be recognized as first among equals, not only because their deleterious effects were so viscerally felt by consumers in the form of raised tariffs and extended periods of power rationing (see next chapter), but also, for the analyst, because of the revealing way such scandals exploited the changing material affordances of electric generation itself. Actual emergency power plants, as a rule, run on imported diesel and heavy fuel oil, rather than the hydro-pressure of national dams constructed throughout the twentieth century. The financialized, "just-in-time" nature of these privately contracted thermal sources (see, e.g., Günel in Rahier et al. 2017) makes them especially amenable to patrimonial rent-seeking and elite accumulation. A growing "energopolitical" scholarship (Boyer 2014) has explored how fuel sources have afforded certain democratic possibilities and foreclosed others.[7] To that effect, a more specific goal of this chapter is to trace connections between Tanzania's shifting fuel mix and its increasingly liberalized rent-seeking across three postsocialist decades, without sinking into a reductive materialist determinism. This in turn requires understanding how the intertwined projects of electrification and nation-building developed through the colonial and socialist eras.

dam nation

Illustrating Susan Leigh Starr's (1999) dictum that infrastructures grow along existing pathways, the complexion of the electrical grid in what is now Tanzania tracks the colonial railway lines that in turn overlaid the older Swahili caravan routes of the late nineteenth century (map 1.1). Power generation originally comprised a handful of diesel generators in the German quarters of Dar es Salaam, the rural district headquarters, connected via webs of rail mostly clustered along the northeast corridor. After World War I, the British inherited the colony and in 1936 built a hydropower station along the Pangani Falls to power its sisal plantations, spurred by the "Chamberlainite idea of extracting value from the Empire's 'great estates' for the British metropolis" (Van der Straeten 2015a, 157). The colonial government formed a licensed concession, Tanesco, to run the railroad-plantation complex, with all shares owned by East African Power and Light of Kenya (EAP&L). Another concession,

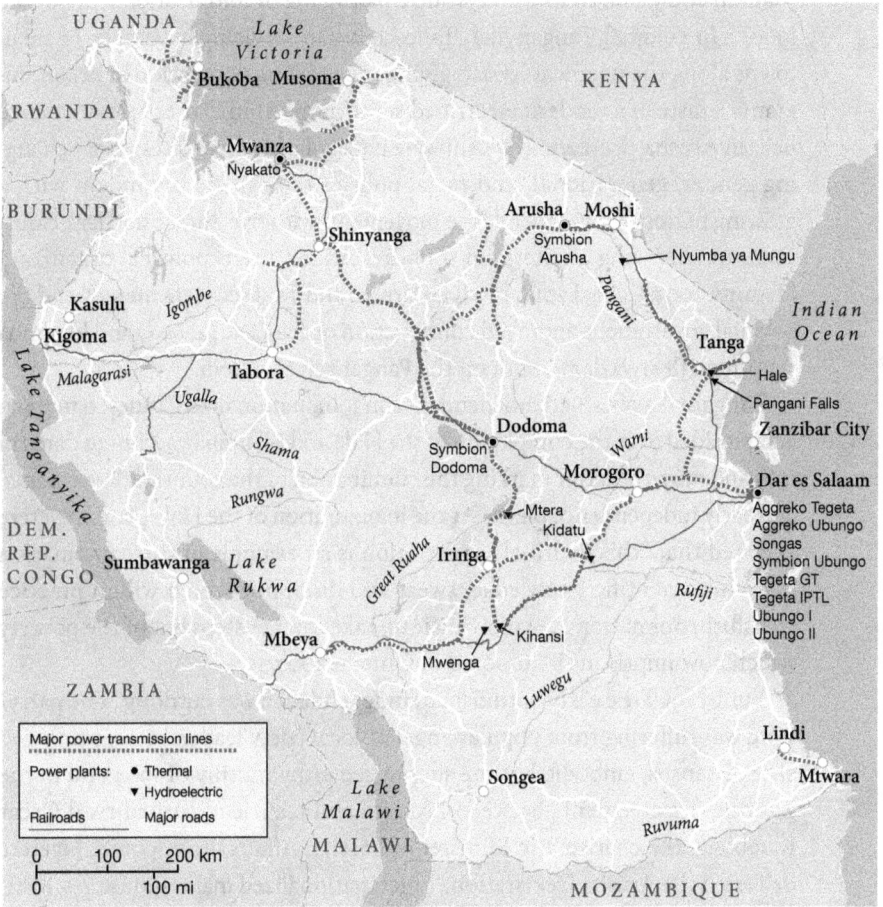

MAP 1.1 The National Grid, ca. 2015.

Daresco, of which EAP&L owned a minority share, was responsible for Dar es Salaam, supplying its government administrative centers and segregated white neighborhoods (Ghanadan 2008, 55). As elsewhere in the first decades of the twentieth century, electricity in East Africa was harnessed to a racialized, extractivist mode of colonial rule that could hardly imagine "natives" as consumers or beneficiaries of electricity (Shamir 2018).

If this "enclave electrification" (Hasenöhrl 2018) mapped out patterns of segregation and extraction, its evolution in the late colonial era (1945–1961) was defined by a new expansive social emphasis on African "development," animated by a growing international consensus around modernization theory

(Rankin 2009) and in some ways anticipating the decolonization wave of the 1960s.[8] In colonial Tanganyika, Tanesco began servicing a number of small towns along the northeast coast, giving the growing population of urban migrants a taste of a modern electrified sensorium. As in Dar es Salaam, dance halls in Arusha, Tanga, and Moshi were particularly potent sites where changing gender, generational, and racial politics took shape in tandem with a growing nationalist independence movement, with live Taraab and jazz bands sometimes serving as sites and soundtracks for TANU political organizing (Tsuruta 2003; Callaci 2011). In 1957, Daresco and Tanesco were merged and the colonial government approved construction of the Hale (21 MW) and Nyumba ya Mungu (8 MW) dams, both on the Pangani river system.

Tanganyika obtained independence in 1961, nationalized Tanesco in 1964, and presided over the completion of the Hale and Nyumba ya Mungu dams in 1965 and 1967, the latter marking the culmination of the country's late colonial and early independence phase. At the inauguration of the Hale plant, Nyerere declared that "this hydro-electric station is an example of the combination of brains, scientific knowledge, sweat and discipline which will in practice transform our nation . . . [allowing it to] take further steps out of the poverty which now imprisons it" (quoted in Öhman 2007, 157).

But by 1967 the early optimism of independence was curdling. The party/state was suffering from popular anger at local party leaders' corruption and elite dissensus, embodied in the suppressed army mutiny of 1964 and in the growing disagreements between Nyerere and his friend-turned-rival Oscar Kambona, the charismatic Minister of Foreign Affairs. In response, Nyerere delivered the Arusha Declaration, which nationalized major industries, instituted a leadership code that prohibited private accumulation by party cadres, and elaborated a nested party structure spanning a national committee down to ten-house cells in each village. The militant party/state policed the dress of urban women, expropriated property from (often Asian) rentiers, and outlined the nation's capitalist "enemies" (*maadui*) and "parasites" (*wanyonyaji*), drawing on a mixture of international and regional idioms of exploitation, as well as a Cold War paranoia about "foreign sabotage" more broadly (Lal 2015, 69). Finally, inspired by growing diplomatic and financial ties to China, it also embarked on an ambitious plan of collectivization that by 1976 had, often by force, resettled over 10 million peasants across the countryside into permanent "ujamaa villages" (Shao 1986; Lal 2015).

Tanzania's socialist period was intertwined with its "big dam era" (Öhman 2007, 40). The late 1960s saw the inauguration of the Kidatu (200 MW) and

CHAPTER ONE

Mtera (80 MW) dams along the Great Ruaha River, both fully operational around 1980, and accounting for the bulk of the nation's hydropower. Like the new socialist capital of Dodoma, the Great Ruaha is located in the middle of the country, further marking a shift away from the colonial coast, as well as a commitment to political centralization and rurally oriented development. Villagization, notably, was partly justified as a logistical move to rationalize mass access to state services like electrification (Van der Straeten 2015a, 275). Liberal international development organizations remained key supporters of this nonaligned socialist developmentalism (Jennings 2002).[9] Both Kidatu and Mtera were funded with assistance from the Swedish Development Association (SIDA) and the World Bank, and partially inspired by a high-level state trip to the United States to see the Tennessee Valley Authority (see Scott 2006) and other mega dam projects (Öhman 2007, 157).

Historians of hydropower in postcolonial Africa have rightly emphasized their uneven developmental effects, exemplified in the irony of high-tension wires towering above unelectrified villages, transmitting current to far-flung urban centers (Tischler 2014; Isaacman 2005). Tanzania's big dams were no exception. The erection of the Kidatu-Mtera system was accompanied by prohibitions on using any upstream waters for fishing or agriculture, as well as forced displacement downstream (Öhman 2007). Rural access to electricity, moreover, has remained miniscule, hovering around 1 percent by 2000 (Ghanadan 2008, 62). In these respects, hydropower joins villagization and its late colonial predecessors as failed exercises in high modernist authoritarianism (Scott 1998; Schneider 2004).

Still, if hydropower did not quite deliver the substance of rural socialist development, it did help establish—and reflexively highlight—the social form of that development. At the inauguration of the Kidatu dam construction in 1975, school children sang that the power plant would "spread so much light over Tanzania that the opponents of socialism would become visible" (quoted in Öhman 2007, 132). Like exchanging a natal village for collective settlements, a mother tongue for Kiswahili, or forgoing capitalist investment for international admiration and Pan-Africanist solidarity, dams were signs of national "self-reliance" (*kujitegemea*), orienting Tanzanians to a collective future. To be sure, these feats of high modernist ordering and the sacrifices they entailed were often resented and inflected with local interpretations (Lal 2015, 45–55). Nevertheless, as chapter 2 will show, they helped impart an ethos of political unity and stability that, however ruefully, could weather the crises of the 1980s and beyond.

By the late 1970s, the massive expansion and centralization of state power that underwrote the Arusha Declaration was overheating. Over the previous decade the government had created over four hundred parastatal companies, pumped massive investment into its small and unsuccessful industrial base, financed the transfer of state ministries to the new capital of Dodoma, supported socialist FRELIMO forces in Mozambique and regional Southern African antiapartheid struggles more broadly, and embarked on an expensive and disruptive war to depose Idi Amin in Uganda to the north. Financing came from foreign aid and agricultural exports like coffee and sisal, a classic "open economy" sensitive to perturbations in the world market (Guyer 2004, 117). After a series of external shocks, state marketing boards reduced prices paid to smallholder agricultural producers, who in turn smuggled their crops into neighboring countries to sell, or who simply "depeasantized" and migrated to cities like Dar es Salaam.[10] In cities, inflation and consumer shortage spawned "parallel markets" for consumer commodities (Bagachwa and Naho 1995), often controlled by those with high-level positions in state companies or the civil service. In 1983, the state responded with a massive operation against "economic saboteurs," rounding up over four thousand mostly Asian small shopkeepers and low-level civil servants (as well as hustlers like Ally), after which "the few necessities like sugar and soap that many people could manage earlier to get on the black market completely disappeared" (Nagar 1996, 72). One well-to-do middle-aged man laughingly described to me how, at the time, his schoolmates would all stand around in a circle sharing a single cigarette, sucked down to the filter.

These crises also fissured the centralized party state. Led by second President Ali Hassan Mwinyi (1985–1995), a wing of reformists favored signing on to the International Monetary Fund's program of market liberalization.[11] Slowly they prevailed, and over the course of his term, they devalued currency, lifted controls for consumer and agricultural imports, cut support for school and medical care, created tax breaks for exporters and prospective investors, and legalized private ownership of parastatals (Aminzade 2013, 253–55). In 1991, the Zanzibar Declaration revised the party's stringent leadership code, allowing cadres to hold second incomes, rent out houses, and hold shares in private enterprise.

My friend Ally nicely captured the ambivalent results of this era when he explained to me that *uwazi* (literally openness) was like rolling down the window of his dilapidated taxi. Yes, you let in some cooling breeze—but you're also liable to get a mouthful of bugs and dust. At first glance there was an obvious pro-social thrust to liberalization. Lifted controls deflated many parallel markets for food and basic necessities. The widespread availability of *mitumba*, second-hand clothes, for instance, was broadly popular, as was the easing of onerous licensing regimes for business. Urban salary-earners such as teachers and clerks had long come to rely on informal hustles, selling food or raising poultry at home, or using state-issued vehicles as private taxis. From this perspective, liberalization merely ratified citizens' ongoing noncompliance with state control. Their activities embodied a "right to subsistence" (Tripp 1997) that in some ways aligned with the basic distributive politics of socialism: everyone is entitled to a cut of collective wealth circulation. "Rolling down the window" oxygenated urban survival.

But it also meant more bugs and dust coming in on the wind. Reform was not only driven by popular noncompliance but by evolving strategies of elite accumulation. As Michael Lofchie observes (2014, 42),

> At first public sector officials participating in the parallel economy had a stake in the preservation of the official economy because the scarcities it created raised the prices they could charge in their parallel market activities. As this stratum of officials developed into a more robust entrepreneurial class, however, [the parallel market] became a constraint . . . many members of the political-economic oligarchy began to understand that they could accumulate greater wealth by extending their profitable activities to economic areas still dominated by the state.

As elsewhere in the world, the first decade of "market fundamentalist" neoliberalism (re)constituted a political-business class, sustained by "rampant tax evasion, private looting for public resources, and the destruction of major local industries" out on the veranda (Aminzade 2013, 255). There was also a specifically racialized dynamic of "extraversion"—Bayart's (1993) term for the *longue durée* practice of "African" (i.e., black) elites acting as gatekeepers connecting the continent to the wider flows of regional and world economies. Amongst the new political-economic oligarchy, "coastal peoples and Zanzibaris with links to India and the Middle East, together with Tanzanian Asians, were well represented" (Kelsall 2002, 610), stoking lingering animosities against racialized minorities who were formally embraced but tacitly persecuted during the socialist era (Brennan 2012).

Not all elites supported liberalization. Managers of parastatals like Tanesco, for instance, preferred their monopoly privileges, as we will see. But there were ideological reasons as well. Nyerere, who vociferously opposed IMF reform in the early 1980s, remained chair of CCM's central committee and consolidated a faction of traditionalists who found ways to slow down the pace and extent of reforms. By the 1990s, Nyerere had become an outspoken advocate for multiparty democracy. This may not be as surprising as it seems. Democratic elections *between* CCM candidates had always been popular and robust, and helped "maintain the premise that legitimate authority was based on the rule of law and not the personal rule of an individual or small elite group" (Lofchie 2014, 3). Along these lines, Nyerere was convinced that multiparty reform could work to rediscipline his party's more entrepreneurial members by making them accountable to mass support.[12] While most Tanzanians viewed multipartyism suspiciously, it nevertheless aligned with changing post–Cold War international norms and did appeal to some segments of Tanzanian society that had suffered in the liberalizing era. In 1992, along with other liberal measures such as media deregulation and freedom of the press, CCM approved multiparty democracy.

In the first multiparty election of 1995, the presidency went to one Benjamin Mkapa (1995–2005). Partly through Nyerere's influence, Mkapa had received the CCM nomination over other popular figures such as Edward Lowassa, whom we will meet below and whom Nyerere judged to have enriched himself too fast in the Mwinyi years.[13] Sometimes known as Mr. Clean, Mkapa ran on an anticorruption platform, promising to uproot the influence of oligarchic *vigogo* (literally, large tree trunks—or more figuratively, "dead wood"; see Phillips 2010). Less to Nyerere's liking, Mkapa also worked to repair a strained relation with donors who increasingly felt that their reforms were being undermined and their aid was being used to compensate for—and thus ultimately subsidize—malfeasance.

With its mix of market discipline and anticorruption, Mkapa's platform resonated with an ascendant donor "politics of the air conditioner" that attempted to extend the reforms of the 1980s while reining in its veranda-style excesses. Along these lines, Mkapa partnered with donors to introduce efficiencies into public administration through direct budget support and various auditing regimes, strengthening certain state capacities like tax collection and border control (Gray 2015, 2). Here, indeed, a certain Foucauldian account of neoliberalism as a style of reasoning comes into view, one interested in institutional regulation and committed, however implicitly, to the reality of a social state.

In key respects, however, "the air-conditioner remain[ed] a mere façade, screening the outside world from the humidity of patrimonial politics" (Kelsall 2002, 598). Some of its shortcomings were owed to the inherent limits of any triangulating centrism. Western donors and NGOs in Tanzania were enamored with the abstract and patronizing language of good governance and civil society, giving rise to a veritable "workshopocracy" (Kelsall, 2002, 604) that, ironically enough, replicated the hierarchical and formalist governance structures of the socialist era, and belied the legalized incursions and dispossessions of Western corporations—notably in mining (Green 2010; cf. Appel 2019, 255–58). Air-conditioner politics were also limited because veranda-style dealings only accelerated with the advent of multipartyism in the 1990s, as wealthy businessmen began donating to political campaigns (Aminzade 2013, 322). For elites less regimented by a centralized party-state, the entrepreneurial freedom to scrape rents was, increasingly, a necessity in order to remain competitive. The interplay of these air-conditioner and veranda-style neoliberalisms, and the creative ways each drew different socialist themes of moral rectitude and national sovereignty, respectively, is no better seen than in the subsequent evolution of power sector reform.

a tale of two power plants

In 1992, the water level in Tanzania's Mtera Dam reservoir plummeted. Worryingly, the Great Ruaha River, the dam's main tributary, had briefly dried up, and Dar es Salaam was forced into daily power rationing for most of the next two years. The year 1994 also coincided with the first run-up to multiparty elections, making the cuts both economically damaging and politically embarrassing. President Mwinyi's Minister of Water, Energy and Minerals, the future President Jakaya Kikwete, blamed the low reservoir levels at Mtera on drought and environmental degradations occurring upstream of the Great Ruaha. In truth, the wet season had quite reliably ensured water in the reservoir; the real culprit for the electricity shortages was more likely poor planning, with Tanesco engineers letting out more water than the dam was designed for to meet growing demand (Walsh 2012). Nevertheless, low seasonal rainfall has indeed created complicated variables for hydropower planning since the early 1990s, and the system has suffered periodic water shortages up to the present day. As prelude or pretext, insufficient hydropower has been crucial to CCM's incorporation of unexpected emergency measures into grid management. The onset of multiparty elections and the Mtera "drought" pushed government players into a reactive, crisis-driven version of what was originally a more cautious energy reform process—a process and narrative of crisis open to political exploitation.

At first, the response to the power crisis was in step with donor orthodoxy. By the early 1990s, electricity service had deteriorated throughout the country, prompted first by the crises of the late 1970s and then worsened by the structural adjustments of the Mwinyi years.[14]

For the international development community, this was a familiar story. A small but influential community of private consultants and think tanks was concerned with low rates of access, technical losses, and underinvestment across the globe (Gratwick and Eberhard 2008, 3949; Wamukonya 2003).[15] The poor performance, along with a combination of improvements in communications technology and the emergence of modular, less capital-intensive generation systems, offered plausible grounds to challenge the theory that utilities were natural monopolies, and thus best organized as vertically integrated state-owned sectors. Over the next decade, drawing on experiences in neoliberal strongholds such as England and Chile, a new approach to power sector reform became codified in a "standard model." Its elements included the following: corporatization (turning a state ministry or asset into a separate legal entity); commercialization (adoption of cost-recovery principles); unbundling (separating out generation, transmission, distribution, and/or splitting each level into multiple entities); establishing regulators and legal frameworks; divestiture (allowing private investors to purchase stock); and wholesale and/or retail market competition.

While the standard model had roots in the structural adjustments of the 1980s, it was ultimately a different sort of creature. While some consultants and policymakers invoked the standard model in terms of a teleology that would inevitably culminate in full privatization, many others provided qualifications, exceptions, and disagreements over the ordering of these steps, the empirical bases for their efficacy, and the need to account for local political context as well as specificities of the existing setup. In both form and content, this was policy not as shock treatment but as *process* (Collier 2011, 159). Just as Collier describes for the infrastructural reform in the Soviet Union in the 1990s (2011, 159), Tanesco was not to be privatized in the "stroke of a pen" sense of sale to the highest bidder (what Nimi Mweta in the epigraph calls the *liquidation* of parastatals) but, ideally, was to be reconfigured stepwise. Finally, as with air-conditioner politics more generally, market logics subsumed but were not entirely divorced from "moral" questions of pro-poor development in the form of their promise to reduce rationing and improve service.

By the middle of the 1994 drought, in line with the growing consensus around power sector reform, MEM authorized emergency measures to attract private investment into the sector, a goal recently affirmed in a new national energy policy. The World Bank provided emergency funds for 75 MW of emergency power. The following year, the World Bank formalized power sector reforms as the condition of continued support with its $200 million Power VI project. By 1997, Tanesco was formally under the authority of the World Bank–financed Parastatal Reform Commission (PSRC), which mandated eventual unbundling and privatization with 70 percent of shares of sales to private investors (Ghanadan 2008, 71).

The centerpiece of World Bank power reform was Songas, a $300 million joint venture financed by the Swedish International Development Cooperation Agency, European Investment Bank, the International Development Association, and the government of Tanzania. It involved constructing a drilling, processing, and pipeline complex from the Songo gas fields up to a facility constructed in the Ubungo area of Dar es Salaam, which would house six gas-powered turbines producing 190 MW of electricity. While Tanesco owned a minor stake (1 percent), the twenty-year power purchase agreement (PPA) was relatively unfavorable for Tanesco, which purchased Kwh at 1.1 cents and sold them to consumers for .075 cents (Wamukonya 2003, 11).

Nevertheless, Songas was a better deal than the alternative, veranda-style proposal. By 1997, Songas negotiations had lagged as state officials diverted their attention to IPTL, a joint venture between Malaysian company Mechmar and 30 percent minority Tanzanian shareholder VIPEM. In 1994 the MEM had signed a memorandum of understanding for emergency power to relieve the drought-induced crisis. In the first of what came to be the proverbial nine lives of IPTL, the company lost out to the World Bank's support for the two 75 MW turbines. However, over 1995 and 1996, IPTL renegotiated a long-term, twenty-year PPA for 100 MW of diesel generation. The agreement was presented to the public in 1997 as a way to meet rising long-term demand, though its urgency was seemingly underscored by drought-induced power rationing in September of that year.

Why was IPTL offered an emergency power contract and then a long-term PPA in the midst of an ongoing, multimillion-dollar Songas deal? Some commentators feel that "although it is easy in hindsight to accuse stakeholders of acting imprudently in the face of emergencies, the actual conditions of load-shedding and shortages"—that is, the droughts of 1993, 1994, and 1997—"appear to have provided few alternatives" to IPTL (Eberhard and Gratwick 2011, 25). And yet the IPTL solution was questionable on a number of counts. In particular, the cost at which it would sell power to Tanesco was to be calculated *after* a determination of IPTL's capital costs—this accurately foreshadowed a cripplingly expensive arrangement that would have significant consequences down the line. More generally, the very structure of the IPTL deal was "a long-term solution to a short-term problem" (Cooksey 2002, 61).

First, a largely Scandinavian-funded hydropower project (the Kihansi dam) that was already underway was expected to come online in 2000 and mitigate further shortages.[16] Moreover, to underline the haste of IPTL, late 1997 and 1998 saw the El Niño storm system refill Tanesco's hydro-network to capacity. Finally, and most egregiously, Tanesco was already in possession of four kerosene-powered generators that were not in use.[17] It would seem, to borrow a phrase from my interlocutors, that "politics had entered the energy sector" (*siasa kaingia sekta ya nishati*)—the often-racialized and factionalized politics of liberalization.

generating race

Who then was behind IPTL? The company's minority shareholders, VIPEM Ltd., comprised a group of Tanzanian Asian businessmen with experience in the Tanzanian Harbor Authority. Their "frontman" was James Rugelamira, a former Bank of Tanzania employee who, through advisory roles as a private sector representative of "indigenous entrepreneurs," had ties to the Mkapa presidential administration. The project found significant support among the political elite in CCM, including the ministers of planning, of finance, and of water, energy, and minerals (the aforementioned Kikwete).

Taking office in 1995, Mkapa was not necessarily convinced, and much of IPTL's drama revolves around the way shifting networks tried to steer him one way or the other. In accordance with his air-conditioner platform, his new cabinet initially "excluded political heavyweights from the previous regime" of President Mwinyi (Kelsall 2002, 606). There was also an ethno-religious dimension to Mkapa's tenure insofar as he was a Christian from the south of the country, succeeding a coastal Muslim president. Nevertheless, because Mpaka required the support of such *vigogo* heavyweights for his 2000 reelection campaign, they remained an important factor in his political calculus.[18] Indeed, it was "precisely the Mwinyi heritage surrounding IPTL that Mkapa [struggled] to contain since coming to power in 1995" (Cooksey 2002, 61). Caught between the force and ambition of the Mwinyi-era reformists on the veranda and the good favor of the air-conditioned donor community that was supplying large amounts of funding to the government, Mkapa maintained an ambiguous attitude toward IPTL. Donor attitudes, by contrast, were unambiguous. The World Bank suspended the Songas contract in protest, and the newly independent/opposition press published a good deal of criticism. Reporting on the pro-IPTL alliance, one foreign Reuters journalist offered the observation that "Tanzanians are possibly the worst exploiters of Tanzanians" (quoted in Cooksey 2002, 63), ironically invoking one of the central political concepts—exploitation—of the socialist era.

Or perhaps not so ironically. While anthropologists have developed a critique of neoliberalism as a "demoralized," technocratic program (Ferguson 2006, 71), popular Tanzanian ideas about socialism and liberalization were in practice morally entwined. Thus, it is not much of a stretch to see how the language of good governance and pro-poor development would find an elective affinity with the socialist discourse of "exploitation." As Todd Sanders (2008, 116) emphasizes, neoliberalism in the 1990s carried many of the same moral valences as socialism, including a commitment to fight poverty and produce an accountable government capable of providing basic services to its citizens.[19] This was indeed the central tension of Mkapa's presidency, which embraced air-conditioner reform as itself a disciplinary, moral corrective to the more free-wheeling liberties of Mwinyi-era liberalization.

At the same time, air-conditioner politics, with its conditionalities and neocolonial aura of dependence on white donors, was anathema in obvious ways to the postcolonial and socialist themes of self-reliance that Nyerere and CCM had consistently promulgated. Supporters of IPTL took advantage of this disjuncture and launched a rhetorical counteroffensive in the press, dismissing criticism as opposition to "South-South" cooperation and a conspiracy on the part of "foreign agents"; that is, the World Bank and Songas. As one of the most vocal critics of IPTL, Cooksey was especially singled out. "I soon ran foul of IPTL. In a long letter to *The African* (1 June 1998) titled 'Brian's phobia against South-South Commission,' I am dubbed a racist with a 'pathological hatred of South-South cooperation,' with 'derogatory tendencies towards African governments, leaders and its people' . . . I am [elsewhere] referred to as a 'dangerous underground advisor,' 'an academic and business crook,' with 'prejudices against African and Asian leaders.' I am further described as a tax-evader, an unlicensed gemstones dealer, a frequenter of a disreputable Dar es Salaam dance hall, a 'foreigner and a self-appointed energy expert'" (Cooksey 2002, 64).

I would suggest that Cooksey's whiteness served as a rhetorical displacement of anti-Asian/anticorruption discourse that had put the Mwinyi-associated "heavyweights" on the back foot ideologically. It aligns both "Asian" and "African" Tanzanian nationals within a broader South-South ecumene by positioning them against the neocolonial interference of the World Bank and its supposed lackeys, embodied here in the figure of the louche expat. Discursively, the dismissal of Cooksey is a complex one. It ties together familiar socialist themes of "friendship," racial egalitarianism, and anti-Western self-reliance.

On the other hand, instead of invoking these deals in the name of the *wananchi* (citizens) as CCM historically had done, the new victimized protagonists in this rhetoric are Asian/African businessmen and "leaders." In this way, IPTL supporters managed to square the postsocialist circle, deflecting one kind of political-racial resentment ("Asian" mercantile capitalism) by invoking another (white neocolonialism).

This dueling rhetoric came to a head over a technical detail. Eventually it was revealed that IPTL had surreptitiously switched up plans for construction, ordering cheaper, medium-speed turbines over slow-speed ones. With no documentary evidence of due diligence about actual project capital costs, Tanesco, with President Mkapa's reluctant blessing, issued a notice of default. The case was brought to The International Centre for the Settlement of Investment Disputes. Bribery to push through IPTL was alleged though never proven. In 2001, after a lengthy arbitration, the case was ruled in Tanesco's favor as project costs and the resulting tariff calculations were reduced.

It was a Pyrrhic victory—IPTL was to still sell Tanesco power at a grossly inflated 13 cents/unit plus an additional capacity charge of $2.8 million/month, whether or not any power was actually generated. By contrast, Songas negotiated a capacity charge that was far less and would decrease to zero by the end of the contract period.[20] The difference in costs for energy actually consumed ("variable energy costs") was wide as well, owing to the fact that IPTL uses expensive diesel. Overall, IPTL was the costliest IPP in Africa: $20 million from the national budget was allocated to bail out Tanesco. Moreover, during the drawn-out IPTL arbitration, the World Bank delayed the Songas project in protest. Over this period, the interest rate on Songas's allowance for funds used during construction (AFUDC) grew substantially, further running up Tanesco debt (Gratwick, Ghanadan, and Eberhard 2006).

IPTL came online in 2002; Songas in 2004. Together they cut the predominantly state-owned, hydropower basis of generation by roughly half. By 2005, the two power plants represented a significant transformation of the national grid. Thermal sources (natural gas and diesel) would now out supply hydropower, comprising over 50 percent of all power produced. Financially, Tanesco continues to operate at a loss, paying an average of $13 million per month for the plants' fuel and capacity charges, well over half of its revenue (Gratwick, Ghanadan, and Eberhard 2006). In turn, the financial strain on power generation has ramified down to negatively affect the commercialization of electricity distribution. In the next section, I discuss this dimension of the reform process.

the problems of netgroup solutions

At the turn of the millennium, the stalled Songas project and the arbitration for IPTL cast significant doubt that Tanesco would be able to follow any comprehensive model of privatization. Thanks to the IPP quagmires, deteriorating infrastructure, and debt owed to Tanesco by government ministries, Tanesco's operational and financial situation was declining. Nevertheless, with donor urging, the government accepted bids for a temporary management contract, eventually won by a small South African company, NetGroup Solutions.

The "temporary" terms of private management contracts were ambiguous. Private management contracts were consistent with developing world utility trends in the 2000s, which saw a retreat from heroic privatization agendas and the embrace of more modest, "flexible" governance instruments (Harris 2003). On the other hand, despite its short-term flexibility, the NetGroup contract was theoretically the *prelude* to full privatization. The idea was that NetGroup would act as a kind of onramp, helping Tanesco to improve its fiscal and technical operations in order to make it attractive to investors and prepare it for eventual unbundling. These goals were pegged to two contract periods, respectively. From 2002 to 2004, NetGroup was meant to improve Tanesco's fiscal situation, and a renegotiated contract from 2004 to 2006 would focus on technical improvements and investments. The tenure of Net-Group over these two periods further illustrates the racialized push and pull of air-conditioned and veranda politics.

In the first contract period, NetGroup was extraordinarily successful, doubling monthly revenue collection from $10–$12 million to $22–$24 million (Ghanadan 2009, 415). These gains can be attributed to a classic austerity package (Ghanadan and Eberhard 2007): tariffs were hiked by a total of 28 percent, while socialist-era cross-subsidies were reversed. In addition, NetGroup raised tariffs for residential and light commercial consumers by 39 percent while lowering industrial tariffs by as much as 28 percent. The residential "lifeline" subsidy for poor households was cut from 100 to 50 kilowatt hours per month. Residential electricity bills tripled overall by 2005. Moreover, small-scale residential and commercial clients bore the brunt of an aggressive disconnection campaign, while middle- and upper-class domestic consumers were serviced with sophisticated prepaid metering systems also imported from South Africa. Finally, NetGroup negotiated a voluntary retrenchment of 1,753 Tanesco workers, or 21 percent of the staff (Ghanadan 2009, 413). In chapters 3 and 4 we will see the consequences of these policies for the street-level experience of electricity.

Besides revenue, however, by any other metrics Tanesco's health was poor. NetGroup oversaw no appreciable gains in new connection rates, system reliability, or reduction in power losses. In addition, one of the most contentious features of the contract was that managers' performance bonuses were calculated solely on the basis of revenue increases, rather than in concert with these other metrics (Ghanadan 2009, 420). Arguably, this policy incentivized management to pursue primitive forms of cost recovery at the expense of customer service, distributional questions, and new connections. These calculations also extended to the fuel supply mix. Tanesco managers described to me how, in 2003, NetGroup allowed the Mtera dam system to drain rather than purchase costly heavy fuel oil from IPTL, which had come online the year before, allegedly damaging the subsystem. The "flexibility" of a private management contract did not lead to creative management of Tanesco—but instead, perhaps, to creative forms of smash-and-grab profiteering.

And yet the NetGroup contract demonstrates the racially inflected ironies and complications of this story. NetGroup was certainly perceived to be a neocolonial venture, part of the white capitalist vanguard of post-apartheid South Africa's bid for regional hegemony (Daniel, Naidoo, and Naidu 2003). In April and May of 2002, Tanesco workers protested NetGroup publicly, with placards bearing President Nyerere's portrait and the slogan "Tanesco, its dams and its electricity are the hard efforts of Nyerere and Tanzanian citizens" (Ghanadan 2009, 412) and, more pointedly, "Boers go home!" (Matiku, Mbwambo, and Kimeme 2011, 198). NetGroup managers had to be escorted into Tanesco headquarters under the protection of Tanzania's Field Force Unit, and utility workers threatened sabotage. President Mkapa addressed the protests by reinscribing the contract within the nation's social contract, emphasizing its technocratic aspects over its racialized political-economic ones: "Contracting a foreign firm to run TANESCO is not only inevitable but also long overdue, and should be understood in the context of state interest to improve TANESCO's efficiency" (quoted in Matiku, Mbwambo, and Kimeme 2011, 198).

There was some popular resonance to Mkapa's claims. Despite its "Boer" provenance, NetGroup also entered a political field structured by other resentments, particularly popular frustration with government corruption and malfeasance. To be sure, NetGroup's disconnection campaigns and price hikes were widely unpopular. Nevertheless, NetGroup won a degree of public approval—eventually, if unintentionally—when they threatened to disconnect the most egregious debtors of all: high-profile government institutions like the police, the military, and even the island of Zanzibar. Which is more perverse, foreign control or a government that (almost) deserves it? The line between these resentments could be confusing, as Rebecca Ghanadan (2008, 88) has observed: "On the one hand, public sector enforcement was welcome if it meant that Mwinyi-era laissez-faire would be replaced with greater public-sector scrutiny and enforcement. However, at the same time, the irony of using *public* monies to pay private managers to carry out *public* collections was not lost on residents who complained local managers could be as 'efficient' if also given Presidential support."

the demise of the standard model

In 2005, the government declined to renew NetGroup's contract. Tanesco itself was "despecified" for full privatization and removed from the oversight of the Parastatal Reform Commission. A key factor in explaining NetGroup's dismissal was the way crippled power generation quickly caught up with the system's new efficiencies. Much of the newly gained revenue was funneled right back out to pay for Songas and IPTL, which had both come online that same year.[21] Against this fiscal drain, the government continued to provide extensive budgetary support to Tanesco.[22] Even the most well-intentioned management, in the most supportive political atmosphere, could not likely surmount these structural difficulties, to say nothing of NetGroup's own complex and awkward presence in the Tanzanian political field.

Still, the decision also reflected the basic illogic of private management contracts (Bayliss 2002), and a certain fatigue with the Washington Consensus in general. Most other experiments in private management, such as the railway concession to the private company REITS and a German firm's management contract for the Dar es Salaam water supply, did not enjoy much practical efficacy either and generally failed to last longer than three years (Bayliss and Fine 2007, 174). In the late 1990s and early 2000s, civil society organizations and opposition MPs such as Tundu Lissu managed to gain political mileage out of negative experiences with other private foreign companies such Barrick Gold (Holterman 2014), while simultaneously championing the causes of "indigenous" black African businessmen (*wazawa*). This was also a time when intensifying diplomatic and financial ties to China provided a counterweight to Western donor pressure and influence.

The 2005 presidential campaign and election of Jakaya Kikwete caught these shifting political winds. Against Mkapa's pro-Western air-conditioner developmentalism, and prompted by the opposition's rhetoric that CCM had abandoned its constituency in favor of "Asian" interests of the kind involved in IPTL, CCM adopted its own version of "rejuvenated radicalism."[23] One source described to me how, upon the cancellation of their contract, the German managers of the municipal water supply were threatened and unceremoniously deported "within 24 hours, leaving their wives and dogs in Dar es Salaam"—a style that echoed earlier deportations and shutdowns of European-run establishments in the socialist era (Heilman 1998, 374; Aminzade 2001, 74). Iconoclastic gestures such as kicking out foreign managers and rejecting Western-style privatization were also a way to signal CCM's nationalist bona fides and shore up its diminishing legitimacy.

While Tanzania's power sector exemplifies "the demise of the standard model" (Gratwick and Eberhard 2008), it has hardly reverted to the developmentalist model of effective (if limited) provision of the 1960s and 1970s. Rather, it has shifted into a hybrid mode in which a public monopoly ironically becomes a key node in overlapping networks of private patronage. Unlike state-owned hotels or breweries that were privatized in the 1980s and 1990s, the power grid is laced throughout the entire economy, its flows widely interconvertible with money, legitimacy, and capacity. Alongside the patriotism of denying Tanesco to foreign investors, it is arguably a richer source of rents as a state asset. We have already seen how this works in IPP/EPP procurement, which honors donor desires to see market reforms while allowing officials to create oligarchic ties with the private sector by disbursing noncompetitive tenders to favored individuals or companies out on the veranda. A similar strategy holds at the distribution level. While the public may have begrudgingly appreciated NetGroup disconnecting various state institutions, "politically condoned nonpayment" (Godinho and Eberhard 2018, 21) and favorable tariff rates for those same institutions is an important patronage tool. As we will see below, Tanesco management itself gives preferential tenders for all manner of operational inputs, from utility poles to business stationery. Because Tanesco enjoys "soft budget constraints" (Kornai 1986), it can usually be bailed out, and thus there is little incentive to address the diminished capacity and associated legitimacy crises that (donors imagine) market discipline would obviate.

That said, in keeping Tanesco publicly owned (and privately useful), the party-state must still contend with its donors, whose continued aid is conditional on the power reform process. It must also contend with a power-consuming public that has benefited from some parts of the reform process (e.g., prepayment) but mostly suffered from it. Hence Tanesco and the MEM have continued to condone tariff increases, austerity measures, and occasional disconnection campaigns (usually of the poor)—but not too much. They will even occasionally revive the full privatization agenda, as they did in 2008 and 2014 (Eberhard and Godhino 2018, 22), only to leave it languishing once again.

Occasionally, however, rentier politicians are too aggressive in goosing the utility, and this institutional equilibrium is thrown off balance. In such moments both donors and the public align in their opposition to grand corruption. But while donors cling to the fantasy of transcending corruption via technocratic good governance, one can detect a more nuanced reasoning, a modal reasoning, about its limits. The *Richmondi* scandal was an illustration of just that principle.

richmondi

In late 2005, amidst NetGroup's aggressive disconnection campaigns and the expense of IPTL and Songas, exceptionally poor rainfall meant that Tanzanians were suffering twelve to eighteen hours of unannounced power outages, prompting NetGroup/Tanesco to formulate a plan for purchasing emergency power, just as the utility had done a decade earlier. Its parent ministry consistently interfered.[24] The MEM informed Tanesco of a number of decisions: it should initiate an international tendering process that would last an extraordinarily short ten days (the standard is forty-five); emergency plants were to be bought, not rented; the criteria for procurement would be made less rigorous; oversight was to be transferred from Tanesco to a three-person government negotiating committee bearing the influence of Kikwete's Prime Minister (and sidelined presidential hopeful) Edward Lowassa. As the parliamentary report on the scandal observed, the selection criteria were "similar to those of a children's game known as *sadakarawe*; 'the winner should win, and the loser should lose'" (Bunge la Tanzania 2007, 34).

In June of 2006, the MEM, through the negotiating committee and with approval from Prime Minister Lowassa, instructed Tanesco to award the tender to Richmond Development Company, an unknown firm supposedly registered in Texas. The signing was reportedly pushed through at night, with the Tanesco head managing director from NetGroup given one hour to review the contract. When the managing director protested, he was "politely reminded that something can happen to [him] if [he doesn't] sign."[25] Needless to say, this "act of humiliation" (Bunge la Tanzania 2007, 45) may have been an indication that the NetGroup contract would not be renewed.

Richmond was contracted to provide 100 MW of diesel-fueled power in the form of five generating plants of no less than 22 MW. The "emergency" terms were lengthy, expensive, and included other sweeteners, such as the provision

that Tanesco be responsible for all taxes involving "importation and expor-
tation of the plant, equipment, tools, spare parts, consumables, testing and
monitoring equipment or in connection with the supplier's performance of
its obligations" (Parliament of Tanzania 2007, 56).

Richmond pressed its luck even further. By September 2006, it had successfully negotiated a letter of credit worth $30 million (35 percent of the total
contract). It stipulated that 17 percent of the total cost of the contract would
be paid once it had submitted documents for the shipment of the first 22
MW plant, and another 17 percent to be paid upon Tanesco's receipt of the
plants in the Ubungo power complex in Dar es Salaam.[26] A month later, on
October 4, Richmond attempted an ex-post-facto amendment, requesting an
advance payment of $10 million before airlifting the first of its plants into the
country, and before providing any documentation that they were capable of
doing so. This was to be followed by a payment of 50 percent of the letter of
credit upon confirmation of the plant's arrival. Tanesco officials protested but
were once again overruled by the MEM. They were instructed to request CRDB
Bank to amend the terms of the letter of credit. The bank did not oblige, citing
the requirement that Tanesco first obtain approval of the Bank of Tanzania and
the Ministry of Finance, which they were pointedly withholding.

Three weeks later, Tanesco informed Richmond that due to its failure to
ship the first plant and begin commercial operations in a timely manner, and
due to suspicions of its actual existence as a registered United States company,
it was in contravention of the contract. After over a month of silence, in early
December, Richmond announced its intention to transfer the contract to another company, Dowans Holdings of South Africa. With more pressure from
MEM, Tanesco agreed to ratify the transfer. Over the next year, Dowans did
manage to deliver the five generator plants, but by that time seasonal rains
had temporarily refilled the hydropower system. Like IPTL, the project was
completed, but belatedly and with much higher core and contingent costs
than the alternative options.

By 2007 the parliamentary report investigation demonstrated that Richmond was nothing more than an unregistered "briefcase company" and
recommended, given its fraudulence and the subsequent conditions of its
transfer, that the Dowans contract be terminated. The generators sat unused
in a warehouse in Dar es Salaam, and Dowans, in protest, took Tanesco to the
International Chamber of Commerce for arbitration. There it was revealed
that its attorney-of-fact was none other than Rostam Aziz, CCM treasurer
and close ally of Lowassa and Kikwete. In 2008, in response to press coverage
and mounting public anger, Kikwete dissolved his cabinet. The MEM minister,

Lowassa, and Aziz all resigned as well. In this respect the Richmond scandal signaled a maturing independent press and a new "Parliament with teeth" (Sitta, Slaa, and Cheyo 2008) that could question—though in important ways only question—the moral legitimacy of CCM. I take up this idea in the following chapter, but I have walked the reader through some of the labyrinthine and legalistic details of the Richmond scandal in order to highlight some other points—in particular, the way it reveals a tacitly modal politics of tolerating exploitation.

emergency power

One hallmark of the *Richmondi* scandal is the way the notion of "emergency" justified modes of primitive accumulation that fell beyond normal governance realms, bearing some similarity to the profiteering logic that was concurrently unfolding in "disaster capitalist" zones like Baghdad and New Orleans (Klein 2007). But how far beyond is beyond? The select committee observed that:

> The major defense made by the leaders of the Ministry of Energy and Minerals is that the country was under urgency and therefore the Ministry had to take the steps that it had taken . . . [but the] Select Committee now understands that their actions were rooted in the bad habit that has become rampant in the country, of people rushing to the accident scene to help when the accident occurs. But not all have the intention to help, to rescue. The intention of the others is to steal, to loot! We leaders who start projects as a result of a national disaster, are not different from the people who rush to the accident scene not as rescuers but as looters. Drought in the country was forecasted, but we did not take appropriate measures early, perhaps we were waiting for a big disaster to occur in order to have an opportunity for projects like that of Richmond. (Bunge la Tanzania 2007, 188)

It is worth taking seriously the committee's rhetoric here. There is a good case to be made that Richmond was a case of truly "grand" corruption, in some sense equivalent to looters who prey on road accidents. However, it is also worth paying attention to an accompanying form of modal reasoning that seeks to *delimit* such corruption, rather than simply condemn it.

For instance, take the committee's consideration of Richmond's repeated, failed attempts to receive advance payment before delivering any of the generators. It became clear in retrospect that Richmond's insistence on advance payment was to buy the generators in the first place since it was a briefcase company with no preexisting capital reserves. Richmond, in other words, was seeking a "loan on soft conditions" (Bunge la Tanzania 2007, 107). We might consider this not too far removed from the kind of "economy of appearances" chronicled so masterfully by Anna Tsing (2000). By working the future anterior temporality of finance capitalism, Richmond tried to bluff its way into something from nothing, simply based on a compelling if largely fictional narrative.[27] To be sure, a failed bluff is effectively indistinguishable from embezzlement. But the select committee was nonetheless moved to imagine the possibility of Richmond, however shoddily, at least delivering the generators on time: "The question that arises here is, if the CRDB agreed to amend the Letter of Credit and pay Richmond Development Company LLC the money it has requested without the approval of the Bank of Tanzania and without the contract allowing for that, would Richmond Development Company LLC have brought the plants into the country as intended or would that have a loss for the Nation?" (Bunge la Tanzania 2007, 101).

In other words, was there a possibility that at least *some* of the value of the deal might have been remitted to the Tanzanian nation in the form of *some* timely emergency power? Richmond arrived at "the crash" and dubiously exploited the situation—but had contingent circumstances not foreclosed the possibility, might it not eventually have been in Richmond's interest to take the injured to the hospital? This is a kind of modal reasoning that acknowledges that a certain amount of exploitation by well-positioned elites may be tolerable insofar as it merely qualifies, and thus preserves, the underlying flow of electricity to the country.

The context that would make some degree of private rent-seeking seem reasonable is of course political-economic liberalization itself. In many ways, "emergency power" was not only a useful pretext for rent-seeking but an apt summation of political strategy in the multiparty era. Cooksey and Kelsall (2011, 35) write:

> Tolerance of damaging rent-seeking deals like IPTL and Richmond can be partly explained by electoral pressures. As political competition within the ruling party and between the ruling and opposition parties increases, so do short-term considerations come to dictate political rent-seeking strategies. One interviewee stated: "[These days] nobody is thinking longer-term." Panic rent-seeking prior to elections allows any initiative to proceed, however shoddy. The same interviewee said: "The political apex cannot stop a dubious deal because nobody is asking and nobody knows what is dubious."

To dismiss (though certainly not to excuse) such scandals as rapacious corruption is to ignore how they emerge from the reality of a political system bending to the pressures of electoral competition. Since the days of single-party rule, CCM has remained in power but its hold is now more tenuous. Liberalization undermines its centralized structure by importing new opportunities for enrichment and influence by ad hoc backroom alliances with *vigogo*/transnational capital. In order to remain competitive, politicians must secure sources of emergency power; they must court and control these new alliances while maintaining some degree of socialist legitimacy.

Whatever the bad faith of this veranda politics, it is one mirrored in an air-conditioner politics that is committed to democratic accountability yet often accompanied by various sorts of foreign dispossessions. Indeed, the parliamentary committee went on to recognize that emergency power profiteering was already an operative mode of power generation that could not be reduced to the excesses of Richmond itself. Richmond was in fact not the only EPP that Tanesco had contracted. Two other foreign companies, Aggreko and Alstom Power Rentals, did successfully supply 40 MW diesel generators for the emergency period. But with their own expensive fuel and capacity charges, these companies were perhaps not much better, the parliamentary committee reflected, and that in some fundamental way "the big burden which TANESCO is carrying under its contract with Richmond Development Company LLC is repeating itself [with these other contracts]" (Bunge la Tanzania 2007, 62). The ultimate problem was not that Richmond had corrupt elements to it, damaging as this was. Richmond illustrated, in an especially exaggerated way, the inherent expense and vulnerability of depending on private power generation in the first place.

In keeping with the structural complicity between air-conditioner and veranda politics, the fallout from *Richmondi* was neither quite a vindication of "good governance" nor, definitively, a triumph of oligarchic impunity. On the one hand, Kikwete reshuffled his ministerial cabinet, signaling that even "the big fish" could face consequences for pursuing their personal interests. The move was hailed as a moment of transparency and accountability by the donor community. Domestically, as with Mkapa and IPTL, it was in line with the sense that CCM would stand up to the corrupting influence of the "Asian" business element (embodied here especially in Rostam Aziz, but also Mohammed Gire, the putative owner of Richmond). As one taxi driver once put it to me: "Kikwete is doing a good job with the country, it's his 'friends' that are the problem." And yet,

> Some critics doubted whether Lowassa's resignation was indeed a victory for the rule of law over the personal interests of high ranking politicians.

Since Kikwete's and Lowassa's presidential campaigns of 2005 had reportedly been extremely costly, it was argued that the Richmond deal may have been finalized to obtain funds to enable not only Lowassa but even the president himself to repay the financiers who had underwritten the campaigns. In this reading, Lowassa's resignation was the result of an agreement between both leaders and their friends to sacrifice one of them to allow the other to remain in office. . . . Whereas the general public understood Lowassa's resignation as a sign of Kikwete's commitment and strength, some analysts viewed it as a clear indication of the weakened position of the president. (Hischler and Hofmeier 2009, 376)

In the next chapter, I discuss how the different experiences of blackouts and power outages mirror this sliding distribution of accountability and opacity, between CCM as the resilient moral torch of the nation or its burnt-out wick. The point is that neither imaginary frame can fully capture the paradigm of emergency power that has flourished in the first decades of twenty-first-century Tanzania.[28] In the current moment, long-term political-economic bases of power and legitimacy are bending and shifting in response to the problems and possibilities of liberalization. This is a dynamic equilibrium characterized by an oscillating *fort-da*, a series of acquiescence and reversals, punctuations and repetitions. Following Marx's ([1852] 2008) disappointed assessment of his own era's repetitions and restorations, we might say that with IPTL and Richmond, emergency power happened first as tragedy and then as farce.

It is also worth pointing out that the "panic" quality of emergency power politics is intertwined with and fueled by a corresponding material energy. That is, in ways that seem to reinforce the short-term horizons of statecraft, the materiality of oil and gas exerts its own effects on the national timescape. By way of contrast, let us first consider hydropower. For hydropower to provide reliable energy, the wet and dry season ecology central to so much of East African life (Evans-Pritchard [1940] 2011) must be accommodated in countercyclical infrastructures—reservoir storage, spinning reserve capacity—that extend to the near future. This foresight is especially necessary in Tanzania, where poor rainfall and rapidly growing demand are already taxing an undercapitalized grid. Oil, by contrast, runs fast and hot. A highly concentrated source of energy, it can be transported swiftly and in large volumes, a testament to the fast capitalism and global supply chains in which the human condition is now suspended (Tsing 2009). But oil's fluid materiality makes for slippery politics. It moves through a capital-intensive infrastructure that labor, civil societies, or national governments find difficult to regulate (Mitchell 2011; Ferguson

2005b). In the OPEC era its price and availability are volatile, contributing to short-sighted geopolitical conflicts. Even the nascent post-carbon imagination is fuzzy on the intermediate mechanics of transition, fixated on visions of green abundance or dystopian resource wars (Harbach 2008). In its own modest way, Tanzania is part of this global timescape in which short-term interest and end-time faith threaten to fold in upon each other. Chained to the IPTL and Richmond contracts, Tanesco is injecting thermal power, and all of its temporal entailments, as a supplement to poor hydrological conditions. Its expense helps foreclose investment in long-term grid improvement, locking in a rhythm of punctuated crises with equally punctuated countermeasures of tariff increases and anticorruption sweeps. Paraphrasing Max Weber ([1930] 2005, 123), we might say that Tanzania's light cloak of emergency power has inadvertently thickened to a hydrocarbon shell.[29]

tanesco management 2007–2015

After the Richmond scandal, Dr. Idris Rashidi, the former managing director of the Bank of Tanzania, was appointed as the new Tanesco managing director. In 2009 Dr. Rashidi proposed to buy the idle Dowans plants outright, since Tanesco was still operating a generation deficit of approximately 250 MW. The move was supported by the new Minister of Energy and Minerals William Ngeleja, but a rival CCM faction denounced the act as unpatriotic and quashed the initiative.[30] In 2011, an American company, Symbion Power, purchased the generators from Dowans. Symbion had already been involved in the power sector as the company contracted by the US government's Millennium Challenge Account to upgrade the transmission and distribution network.[31] Symbion's purchase of the generators coincided with another round of drought and rationing (and the beginning of my fieldwork, see next chapter), for which Symbion was contracted to supply emergency power. After changing ownership multiple times, the Richmond (now Symbion-owned) generators arrived, so to speak. They began generating expensive thermal power with extensive capacity charges to be sold to Tanesco, as originally intended. In July 2013, President Obama visited the Americanized generators onsite in Dar es Salaam to announce his Power Africa Initiative, hailing it as "the kind of public-private partnership we want to replicate all across the continent."[32] The strange career of the Richmond-Dowans-Symbion generators thus highlights the complex entanglements that belie seemingly easy distinctions between a corrupt politics of the veranda and a donor politics of the air conditioner, since the machines have at various times powered each.

Such entanglements are evidenced in the mixed impulses of Tanesco management itself. Before the generators were sold to Symbion, Rashidi played a technocratic Cassandra. He emphasized that the unsavory political provenance of the Dowans generators was irrelevant and warned that unless the Tanzanian government purchased them to compensate for the grid's chronic generation shortfall, the country would again be "plunged into darkness." He also oversaw unpopular petitions for tariff increases, explaining in 2007 that the cost of supply to power consumers was then around 15 cents per kWh, whereas the average tariff charge was around 6.7 cents per kWh. A fully cost-reflective tariff would have to be raised 118 percent but Tanesco had to content itself with 40 percent granted by the newly formed regulatory body EWURA. Tanesco also secured a $240 million loan from private banks and financial institutions to try and kickstart Tanesco's financial solvency.[33] However, Rashidi eventually became embroiled in scandals of his own, involving diverting millions of dollars in government funds for Tanesco to refurbish seven private homes of other top officials in the utility.[34]

This blend of technocracy and graft repeated itself in Idris Ally's successor, William Mhando, who also oversaw tariff increases in roughly the same proportions. As I described at the beginning of this chapter, Tanesco proposed a 155 percent increase in 2011 and was granted a 40 percent increase in 2012. Yet in late 2012, MEM announced that Mhando would be removed due to a suite of mismanagements. First, there were a number of procurement improprieties, including substandard transformer joints and awarding a tender to provide stationery to a company owned by Mhando's wife. Then there was a "ghost LUKU vending system" where a businessman had essentially formed a private distribution company within Tanesco and was selling unofficial meters and kilowatt hours to large power users. Finally, there was a bizarre scheme in which utility poles produced in Iringa were transported to Mombasa and then "imported" with South African labels.[35]

Most interesting, and telling, was the ministry's accusation that Mhando's senior staff had deliberately organized power cuts to pressure the government to underwrite the commercial loan they were seeking to mitigate the costs of yet another emergency power plan in 2011.[36] The conflict soon extended to Parliament, where a number of members of the Parliamentary Committee on Energy and Minerals protested Mhando's suspension and moved to discipline the MEM minister with a vote of no confidence. The ministry fired back, accusing members of the committee of having business interests in supplying goods to Tanesco such as tires and, most damningly, for receiving bribes by oil companies for inflated diesel tenders.[37] In October 2012, Mhando was formally dismissed by the Tanesco board of directors and faced criminal prosecution.[38]

The MEM's post-*Richmondi* crackdown on Tanesco's mismanagement and graft is in some abstract sense admirable, but it more concretely reflects the way anticorruption sweeps help hold air-conditioner and veranda forces in some sort of postsocialist equilibrium. I have discussed a similar dynamic at the ministerial level, in which Kikwete's post-Richmondi cabinet sweeps displaced complex structural causes into a Manichean drama of crime and punishment—"sacrifices" to a postsocialist dispensation larger than their own actions. With Richmond, the balance between backstage rent-seeking and front-stage legitimacy tilted into scandal, and the dismissal of Lowassa was the political concession to stand it upright again.

These same dynamics filter down to mid-level civil servants such as those staffing Tanesco management. With the artificially compressed wages of the structural adjustment era, Tanesco managers (and state civil servants of all kinds) use rent-seeking practices to "pay themselves" and their own networks of influence within the bureaucracy. A key corollary to this idea is that Tanesco management benefits from the parastatal's failed privatization, because if nothing else the flow of emergency power money filters down to them as well. In the post-Richmond and IPTL era, as commentator Nimi Mweta observes, "what [Tanesco managers] want to see is energy being prioritized by the government, and they continue getting the usual Tsh300 bn or Tsh400 bn in subsidies annually."[39] Hence both Rashidi and Mhando sought state-backed loans in the name of keeping Tanesco solvent, and both were involved in significant rent-seeking schemes.

Eventually, however, these two levels collide. Tanesco's ministerial superiors in many ways set the entire system into motion by collecting rents at the generation level (via procurement scams such as IPTL and Richmond), but the consequences catch up with them at the operational level. Given the public relations and financial strain in the post-Richmond era, the MEM was increasingly reluctant to disburse more funds to its parastatal client. One potential solution, as we saw in the opening scene of this chapter, was to raise tariffs—a strategy that both Rashidi and Mhando, in their more technocratic mode, endorsed. Consultants and donors such as the World Bank advocated this approach. But this was politically unpopular and opened the parent ministry to attack from the public and the parliament. Hence a clever solution presents itself: "cleaning up" corruption practices in Tanesco management, just as Kikwete undertook his own ministerial sweeps. Kicking out managers allows the ministry to claim that it is the former's misbehavior, rather than structural underdevelopment and informal privatization, that accounts for Tanesco's difficulties. It also allows them to justify restricting tariff raises below cost-reflective levels (since "mismanagement" is to blame), which would be politically unpopular, and in certain cases to limit budgetary support to Tanesco. The strategy is thus one of partial concessions to the problem of Tanesco's underperformance. A 40 percent (rather than 120 percent or 155 percent) tariff increase every three years, and a new manager every three years: these are modifications—short-term, quick fixes that minimally appease donors who believe, or at least pretend to believe, in the efficacy of air-conditioner solutions such as promoting transparency or good governance, and provide justification to continue to disburse their funds. However, as Mweta candidly observes: "The idea that what is required is transparency in management fails to notice the bottom line, that graft shall simply take other avenues, and if the minister or permanent secretary shuts out all those avenues, there will be a sort of labour unrest, locking out of managers and the rest. The point is that peace and tranquility in public institutions is often woven with graft, where employees engage in informal taxation of their environment and raise the cost of services for their job satisfaction."[40]

Thus, Tanesco managers are in effect given a sort of high-reward but high-risk deal, in which they may head the utility—and parasitically "eat" from it—but are nevertheless always somewhat in the line of fire should the presidential administration, either through pressure from the public, media, donors, opposition parties, or dissenting internal factions, be required to make gestures to good governance and transparency. This is not to say that Tanesco has taken these terms lying down. Recall that in 2012, the MEM accused Mhando of deliberately organizing power cuts, and of using his MP allies/business partners to push through a vote of no-confidence against the MEM minister. To the degree these accusations are accurate, we can indeed think of them as a kind of "labor unrest," a surreptitious strike over the flow of funds to Tanesco and the ability of management to stay in power long enough to profit from them. In chapters 3 and 4, I show how a version of this same labor struggle manifests itself at the street level of underpaid casualized Tanesco workers, who struggle for ways to "pay themselves" and who are thus also vulnerable to moralizing castigation. From the top to the bottom of the state apparatus, we see that its infrastructures may host parasites that exploit them for private gain, so long as they remain within certain tacitly understood limits.

As a result, the compulsiveness with which managers (and even ministers) are ejected from Tanesco's administrative infrastructure parallels the compulsiveness with which emergency power is injected into its material infrastructure. These modal variations act as a substitute to raise cost-reflective tariffs and/or pursue long-term plans for Tanesco reform or overhaul, and thereby allow different sociopolitical actors—the ministerial cabinet, the donors, and parastatal management—to benefit. From an outside perspective, however, the system looks disorganized, as if the government were simply about facing every few years. Wryly invoking the command-economy lexicon of an older socialist ecumene, Mweta reflects that it is as if CCM were no longer in possession of "any bearings as to how problems like power shortages can be resolved, or to speak like the Chinese . . . [possessing] no 'theory' guiding discussion on the issue. . . . Such an outlook existed [i.e. privatization]; it was rejected as unpatriotic, so none is available."[41] The result has been a provisionally managed power sector that is neither wholly private nor public, but an infelicitous—yet surprisingly durable—hybrid of the two, held in dynamic equilibrium.

conclusion

Here we can return to the scene that began this chapter, in which residents of Dar es Salaam found themselves contemplating exactly why Tanesco needed to raise its tariffs when its service was so poor. Was drought the culprit? Underinvestment? Mismanagement? In an over air-conditioned Karimjee Hall, a building that was financed by an Islamic merchant dynasty, that housed the Tanzanian National Assembly, and that now hosts donor and civil society-related conclaves, Tanzanians like Othman Ramadhani confronted these very same legacies as they have played out in the energy sector. Inspired by anthropological scholarship that apprehends neoliberalism as a project of elite accumulation *and* as a style of market rationality, by a historical analysis oriented to long-term continuities, and by an energopolitical sensitivity to material force, this chapter has shown how the flow of current and currency, and the underlying social link between citizen and state that subtends it, have shifted into a strained mode of emergency power, characterized by privatization drives mitigated by nationalization, and rent-seeking networks mitigated by gestures to good governance. In Karimjee Hall that morning, raising tariffs could seem both pragmatic and unreasonable. The last two decades of energy sector reform show how this paradoxical settlement emerged: from drought and donor pressure, the technologies of EPP contracts and thermal generation, the emboldened schemes of a fraying party contending with democratic elections, and resurgent networks of capitalist flow and accumulation. The result is both a social and technical power that operates as a labile "emergency" with no evident "theory guiding discussions" but, upon closer inspection, has generated a surprisingly durable settlement. To capture the full complement of forces that have produced this condition, this chapter has utilized a wide-angle historical and political economic lens. However, I wish to end by briefly zooming back down to the experience of individual consumers like Othman Ramadhani. For many urban dwellers in Dar es Salaam, these histories are encoded in the strain of price increases, poor service, and flickering lights. In order to make sense of these failures, residents not only draw on deliberative debate in the newspapers and halls of Dar es Salaam, but also—as Othman's own frustrated monologue suggests—on an embodied knowledge gained through sensibility and habit. It is to the everyday experience and interpretation of power loss that I now turn.

We, the people of Tanganyika, would like to light a candle and put it on the top of Mount Kilimanjaro which would shine beyond our borders giving hope where there was despair, love where there was hate, and dignity where there was before only humiliation.   —JULIUS NYERERE

the flickering torch   2

POWER AND LOSS AFTER SOCIALISM

In 2013, fifty-two years after national independence, the writer Mlagiri Kopoka pondered the contradictions of life in Dar es Salaam, struggling to square the impressive description of Tanzania with his own suffering of poor infrastructure and economic volatility.

> Certain statistics are startling, if not surprising. The other day, I was watching TV when a certain MP [Member of Parliament] uttered what has taken a long time for me to understand. According to him the economy is growing at seven per cent and he called on us to be proud of such an achievement because our economy is very healthy. He was speaking with such zeal that I started wondering about the different worlds we live in in Bongo; those of four wheels, standby generators and booming economic growth and those of *daladala* [minibus] users, lengthy blackouts and everlasting inflation.[1]

By the early 2010s, persistent power loss in Dar es Salaam undercut the idea that Tanzania was developing, that its economy was growing, and that its citizens should be "proud." As discussed in the last chapter, the affliction of power outages has its roots in the contested, World Bank–sponsored privatization process of the 1990s. Against this backdrop, entrepreneurial elites began taking advantage of drought-induced emergencies, awarding government tenders to well-connected and sometimes dubious private firms for infusions of oil-generated electricity. Such public bailouts were quickly converted into private rents that raised prices, generated scandal, and frayed the public trust. In this way, power generation became a central medium of Tanzania's postsocialist settlement, in which the continuity of a de facto single-party state belied all manner of decentralized, competitive dynamics.

This chapter argues that power loss was a key way in which this postsocialist settlement was experienced by the public. It describes how, over the 2010s, Dar es Salaam residents tracked the power supply's shifting patterns of transmission and distribution and, by extension, interpreted the state of their social contract. Depending on *how* the supply faltered (and how and why it was reported to have faltered), the public might react with rueful solidarity, or suspicion, or resentment. In this way, energy infrastructure became woven into a larger historical reckoning, one connected to the protests and upheavals of the period, that found Tanzanians wondering whether they should continue to support an increasingly repressive CCM, then entering its sixth decade of rule. Flickering lights, I show, became entangled with larger questions of authority, suffering, legitimacy, and violence.

signs and signals

Infrastructures communicate. In a bid to get away from a reductive materialist determinism, anthropologists have emphasized that things like bridges, pipes, and wires do not necessarily sink into the taken-for-granted background of daily life, but are rather seen and sensed in ways that evoke the broader political conditions governing their existence. Leo Coleman (2017) has traced electrification as a political ritual of India, its purpose less to power any particular practical activity than to express the grandeur and modernity of the postcolonial state. Christina Schwenkel (2015) has examined the spectacle of an engineered public fountain in postwar Vietnam, its capacity to contain water partially disabled so as to produce the sensation of overflow, a modernist cup that runneth over. Even when engineered systems are primarily designed to provide functional utility, ritual ceremonies of commissioning or decommissioning become important moments when their aesthetic and symbolic dimensions are temporarily foregrounded (Larkin 2008, 16–18). They hail subjects in a way that is provisionally separate from whether or not a bridge is actually crossed or a service line ever extends out to a household (this, as I suggested in the last chapter, was one effect of Tanzania's ritual inauguration of its "big dams" in the 1960s). At the same time, aesthetics can be foregrounded in more quotidian moments of "ritual" suspension, whenever we notice, say, the smoothness of new asphalt (cf. Khan 2006) or the particular geometry of a building.

This chapter is also concerned with infrastructure's power to address political subjects, but my focus is less on the aesthetic dimensions of infrastructural hardware and more on its everyday function as both sign and substance of the social contract. Public infrastructures, Larkin observes, "represent a relationship between citizen and state. . . . They never just supply water, electricity or gas. They implicate the very definition of the community, its possible futures" (2018, 189). Larkin is getting at the performative element of any social relation based on reciprocity, which cannot be seen but must be inferred through the actions that substantiate it. "Actions speak louder than words." As long as the supply of water, gas, or electricity holds as expected, citizens have some evidence that their state is committed to providing for them. This opens the way for the community to actualize its own commitments in turn, which canonically include recognizing the state's authority, paying taxes, and adhering to norms of civility (Mbembe 2001, 89–94; Mains 2019, 95). The cyclical call and response of these action-flows bring both sets of actors into a social form, with each met obligation contributing to the possibility of the next— their future together remains open.

What happens if and when such vital flows fail to reach their destination? In general, the answer depends on how actors make sense of that failure. Drawing on Kockelman's mixture of semiotics and symbolic-interactionist theory (2017, 175–77), we might sketch this inferential process as follows: I observe how you respond to some input, or behave in some circumstance, and confirm or update my assumptions about you. For example, if we are on a date, and you do something I would not have expected—say, treat the waiter poorly—I try to make sense of the situation. Perhaps the circumstances were not quite what I assumed (e.g., you have a past with this specific waiter and that made your rudeness forgivable in light of the kind person I know you to be), or perhaps *you're* not quite what I assumed (you're a jerk).[2] Having provisionally made sense of the situation (a sense that may be more or less consciously formulated), my responses flow accordingly—say whether I relax or not, whether I agree to another drink. Needless to say, you may be well aware of all this and tracking me just as closely, and/or acting in light of my tracking.

Such signaling extends to economic interactions. Consider Guyer's (2004) ethnography of "waiting for petrol" in Nigeria in the late 1990s. During this time, a sudden fuel shortage threatened to paralyze the life and livelihoods of the entire country, and Guyer accompanied friends to a rural station that had miraculously received a single tanker. The atmosphere, she writes, was "crowded, hot, tense, unpredictable on every score" (2004, 108). And yet, impressively, the station proprietor Madame A kept the scene from escalating. Through a skilled combination of premiums and discounts on waiting and volume, she preserved her right to turn a profit, the rights of buyers to a vital necessity, and the play of difference in local status. In short, there was a rhyme and reason to the distribution. It remained within the "very definition" of buyer and seller, within the "vague whole" (Verran 2007, 181; see introduction) that is the petrol economy in general. The crowd understood that Madame A had not caused the shortage, nor did she exploit it by unduly raising prices and cutting out poorer consumers' access to the market. She remained merely the messenger, responding (qua "output" or "behavior") logically to an upstream problem (qua "input" or "circumstance") that was beyond her. And this itself was a sign, Guyer argues, that the basic social relation, the sense that everyone could play by the rules without having to resort to violence, was holding its form.

Not all difficulties "make sense" so felicitously. The central theorists of moral economy take as their point of departure situations in which urban crowds or rural communities suffer in ways that by their reckoning go *against* the grain of customary notions of justice (Thompson 1971; Scott 1977; Hobsbawm 1969). The resulting mix of backstage anger and everyday resistance against "parasitic" landlords, speculators, or middlemen may die down, or it may well break out into open insurgency. The degree to which ruler and ruled can go on together depends on the extent of maldistribution, the way it is subsequently justified or alleviated, and the strategic advantages and disadvantages each set of actors holds.

In the sections that follow, I explore different kinds of breakdowns in the power supply as signals that the public read and responded to. Recursively, though, such responses usually involved some degree of trying to figure out how and why it broke down in the first place. Unlike the queue at a single petrol station, residents cannot see where electricity is (or isn't) flowing along the grid as a whole, to say nothing of power generation facilities upstream. They must in part rely on authorities to publicly address and explain the misfortune, creating scope for all manner of head games and Goffman-esque impression management. In such circumstances, residents are not only interpreting power outages, they are interpreting what Tanesco or its parent-ministry says—or doesn't say—about them as well. Thus, even if there is an official schedule of load-shedding, suspicions may linger. Daniel Mains (2019), for example, shows how residents of Jimma doubted the Ethiopian state's official explanation that the city's load shedding was required due to low levels in its hydropower system during a robust rainy season. Part of my goal is thus to contribute an ethnographic description of what Eric and Ulrika Trovalla (2015) call "infra-divination"—the way people continuously triangulate information from media sources and rumor with an embodied feel for the starts and stops of current, water, and data flowing through the pipes. Another, though, is to outline how that divination is connected to an underlying moral economy that can tolerate some kinds and degrees of power loss better than others.

two or three ways to lose power

Since the start of the emergency power era in the 1990s, the Tanzanian power supply generally has cycled between two or three main patterns, each characterized by a different configuration of *vital force* and *social form*. In the best of circumstances, residents of Dar es Salaam (where the network is densest) are subject to merely sporadic cuts. Because total installed capacity sits just above peak demand, different areas of the city are temporarily disconnected across the course of the day to keep the system stable. As such, they tend to provoke a "business as usual" response among residents. My own years in Dar es Salaam included unexpected cuts during an evening *Iftar* (breaking the fast during the month of Ramadan); a hip-hop festival; a graduation; a community play; a church service. In almost all cases everyone carried on as best they could, often barely deigning to acknowledge the problem: candles were lit, a flashlight beam tunneled across a mat laden with food to break the Ramadan fast, or actors kept reciting their lines on stage. Most memorably, the church congregation redoubled their singing, jolted by the suddenly tinny quality of their unadorned, de-amplified voices. At this level, there is indeed something to the counterintuitive idea that infrastructures "fail routinely" in the Global South (Edwards 2003, 188). Outages may be sporadic, but they are relatively short, infrequent, and not *too* conspicuously uneven as they spread out across the city; they nibble at the edges of a basic integrity. In this sense the supply has a sufficient degree of both force and form.

Still, the danger here is to assume that urban populations will put up with just anything. At some point, cuts will spike beyond normal thresholds, stretching out to days or nights and/or clustering together with unusual density. Like other kinds of misfortune, they raise questions of causality and responsibility (Evans-Pritchard 1937; Bennet 2005). Alongside the usual rerouting of daily activities, people begin scanning for the kinds of *matangazo* (announcements) that they expect to accompany the outages, and for the order they promise to restore or disclose. Sometimes, the problem is merely technical. Tanesco may announce a temporary shutdown for certain parts of the city in order to maintain or repair a substation, for instance.[3] The darker scenario is that increasing outages are the prelude to an indefinite period of countrywide load-shedding due to problems at the generation level, including during the difficult drought years of 2006 and 2011 when much of the country was without power for six to eighteen hours a day. In these cases, it is the *absence* of power that flows (more or less) evenly and reliably. The same integrity holds, just inverted in a kind of photo negative; the supply has social form, but minimal force behind it.

The ways people responded to collective load shedding raised historically freighted issues of suffering and sacrifice. In many ways, the socialist project turned on just this kind of form without force. Tanzanians may have been poor compared to their "man-eat-man" capitalist neighbors, but they were poor together, united in sacrifice for a better future. During the extreme power deficits of 2010–2011, government and civil society leaders reached back to these themes, interweaving load-shedding schedules with a rhetorical "return to Nyerere" (Fouéré 2014) that posited a unified country combating drought, corruption, and neocolonial underdevelopment. Not all were convinced, as we will see. Looking around at uprisings in Egypt, Algeria, or Senegal, they wondered why a country once again "sitting in darkness" (*kukaa giza*)—so obviously rooted in the rent-seeking of a hollowed-out ruling party—hadn't provoked similar demonstrations or protests. In general though, the state's ability to impart some baseline logistical and narrative order to the power crisis seemed to win it a rueful benefit of the doubt, at least for the moment.

From 2012 onward though, the power supply did not do much to sustain this tenuous benefit of the doubt. As the country moved through the latter half of the Kikwete administration (2005–2015), the supply tended to improve in absolute terms but in some ways became more mysterious. Between ordinary distribution (form with force) and its rationed absence (form without force) was a no-man's land of force without form: a situation in which unplanned outages swelled in frequency, duration, and maldistribution in ways that didn't quite make sense, a feeling amplified by Tanesco's conspicuous inability, or perhaps unwillingness, to explain them. Unlike the older "socialist" style of scheduled load shedding, these swirling outages signaled a fragmenting social contract. Residents divined a kind of "fake rationing" (*mgao feki*) or "silent rationing" (*mgao bubu*), the work of a political class allied with the wealthy and prepared to preserve its prerogatives by fiat. These theories were given credence by the ways arbitrary power cuts echoed and sometimes directly coincided with the state's violent suppression of dissent. The presidency of John Magufuli (2015–2021), which I touch upon in the book's conclusion, would intensify this suppression. This chapter might thus be understood as a prehistory of the Magufuli era, when a postsocialist party state began to oscillate between popular supplication and violent impunity, an oscillation telegraphed in the changing configurations of power loss, its explanations, and its silences.

Finally, a methodological note: the empirical foundation for this chapter comprises my own experience living in Dar es Salaam, as well as roughly three hundred English and Kiswahili language news stories, announcements, blog entries, and social media posts. Many of these sources were read in situ, sometimes on a battery-powered laptop, waiting for the power in my apartment to return, and sometimes at a generator-powered café with a newspaper and cup of coffee. Others were collected from online databases after I returned to the United States. Together, they allowed me to reconstruct citywide patterns of power loss over months and years. But they also constitute a kind of affective archive, tracing the city's emergent moods as it cycled through those different periods of shortages and rationing. Alongside their many practical difficulties, outages generated a whirl of newspaper editorials, speeches, and press conferences, as well as television sketches, tweets, and blog posts. If what follows is an ethnography of power loss, it is also an ethnography of a city that has long been constituted through the public circulation of texts, a history foregrounded in a number of important recent works (Hunter 2017; Callaci 2017; Brownell 2020). The early 2010s, moreover, were a generative time for the production of such texts, as new communication technologies and early social media platforms like Facebook and Twitter arrived in urban Africa and expanded the scope of public discourse. While keeping my feet firmly on ethnographic ground throughout the chapter, I also channel these discursive currents to show how they swirled around the everyday experience of lights flickering on and off, both reflecting and feeding back into a sense of collective experience. To set the scene for that analysis, I begin with a brief overview of power loss during the first half of the Kikwete years, from 2005 to 2010.

"If we care to retrace memory steps," an editorial in the *Citizen* read in 2012, "it [has been] a long time since Tanesco customers experienced uninterrupted power supply. Mostly they remember brief periods of respite, before descending into what starts as intermittent outages, settling into prolonged blackouts."[4] This descent reached its first nadir in late 2005, when a drought crippled Tanzania's hydropower network, which at that point comprised roughly two thirds of the 953 MW capacity system as a whole. As Tanesco tried and ultimately failed to secure emergency power to mitigate the shortage (see chapter 1's discussion of the *Richmondi* scandal), the country phased through periods of six to eighteen hours of official daily power rationing for much of the next year.

The country first began eight-hour daily rationing periods in February of 2006, which were then extended to twelve hours in late March.[5] A slight reprieve held for April and May as the *mavua ya masika* (spring rains) refilled the dams.[6] But by early June, water levels had fallen below the minimum threshold, necessitating extreme twelve-hour cuts daily once again.[7] At this point, Tanesco was forced to rely on the Songas gas-fired plant for nearly half of all power consumed in the country, but its machinery strained under the load. In August, continuously high rotation speeds put excessive pressure on joints of the drive shaft and the plant had to be shut down for repairs, forcing Tanesco back to its dangerously low hydropower sources.[8] By mid-October, Tanesco completely shut down the Mtera and Kidatu dams—the backbone of the hydropower system—which required a complete shutdown of all power to the northern regions of the country linked to the grid.[9] A month later, the rationing schedule was further extended to five more hours in Dar es Salaam, from 6 p.m. to 10 p.m. twice a week, in addition to the weekday 7 a.m. to 6 p.m. schedules, with only government hospital and military installations spared.[10] By mid-December, repair of Songas had eased the Monday through Friday schedule to "almost half a day," though the relief was spiked somewhat by the announcements of increased tariffs.[11]

By 2007, rationing ended as two-year contracts for 340 MW of emergency power from Alstom Power Rentals and Aggreko came online.[12] Despite a number of localized outages throughout the country due to maintenance or breakdown, improved hydrological conditions and this supplementary emergency generation meant that the next few years were relatively stable.[13] The next sizable rationing period occurred from roughly September 30 to October 25, 2009. Another combination of a plant breakdown, a fuel deficit for Songas, and low water levels in the dam systems precipitated an 80 MW shortage. At this point, some members of the government advocated switching on the expensive and controversial Richmond/Dowans and IPTL generators, both lying idle due to legal disputes over ownership and capacity charges, respectively.[14] On October 21, Kikwete ordered the 100 MW IPTL generators to be turned on, at significant cost. In early December, IPTL reduced its production from 70 MW down to 10 MW due to a fuel shortage and a switch in heavy fuel suppliers.[15] This left some areas of the city to experience regular power outages from 7 p.m. to 11 p.m., though Tanesco insisted that there was no load shedding per se, and brought out a line it would repeat often in the future: The outages were instead due to physical breakdown in parts of the aging distribution network.[16] Kikwete won reelection the following year, in November 2010—with the narrowest electoral margins in the country's history, as we will discuss below. At the presidential swearing-in ceremony, the Minister of Energy and Minerals William Ngeleja memorably proclaimed that "the second phase administration of Jakaya Mrisho Kikwete, President of the United Republic of Tanzania, will be remembered by Tanzanians as the period in which the power supply troubles were finished."[17]

The first point to highlight here is just how complex load management is, particularly when installed capacity—the amount of energy the system is capable of providing—sits so close to actual demand. Causality becomes entangled and retroactive; low water levels precipitate emergency plans, which themselves depend on the availability of fuel, financing, and the complex technical process of bringing thermal plants online in a timely and efficient manner. Nevertheless, the "toiling ingenuity" (Guyer 1993a) of load management highlights the efforts made to preserve some sense of systemic coherence that I am calling social form. As the scale of the power loss grows, so too does the importance of practical and narrative containment such that the basic premise of a responsible state is preserved. And yet, over time, no matter how excusable or understandable, the sheer repetition of load shedding surely degrades this precarious achievement. When I began my fieldwork in 2011, smack in the middle of the worst load shedding since 2006, this sense was very much in the air.

2011: annus horribilis

Shortly after Kikwete's swearing-in ceremony in late November 2010, and despite the minister's promise that rationing would be a thing of the past, Tanesco once again announced nationwide rationing[18] of at least three days a week for up to ten hours.[19] As late as February 15, the rationing had increased to as much as twelve hours a day, five days a week; water at the Mtera Dam had fallen to just 1.32 meters above the minimum reservoir level.[20] Lack of sufficient rains in March and April meant continued load shedding, while in May a shutdown of the Songas gas wells would prompt a weeklong surge to sixteen-hour days of power rationing. In early June, the American company Symbion turned on the Richmond/Downs generators it bought in an interim agreement (see chapter 1), adding 120 MW to the grid.[21] While some reports suggested that the rationing eased over the next few weeks of June, others alleged it remained unchanged until Tanesco announced a "new" load-shedding schedule of alternating twelve-hour daily and six-hour nightly national power cuts.[22] Adherence to that schedule, however, was often loose, rendering it "as inscrutable as those Ancient Egypt hieroglyphs."[23] Areas with twelve-hour rationing might stretch out to eighteen hours or more; or in at least one case, simply black out for days on end.[24] After nearly eight months of extensive power rationing, the country had reached a self-described "national disaster,"[25] possibly the "worst power shortage since independence half a century ago."[26] It also directly coincided with my arrival for long-term fieldwork. As in 2006, it would necessitate another emergency power plan: Tanesco contracted Symbion to provide an additional 200 MW of A-1 jet fuel–generated electricity to be installed in 50 MW increments over the next three months, and Aggreko to provide another 100 MW that would come online in late October. Until then I, along with the rest of the city, suffered through twelve- to eighteen-hour outages, and joined up with the public mood as we habituated ourselves to this extreme deficit.

How did the outages feel? In many ways, 2011 simply saw a downward shift in economic activity and general economic prospects. Like oil, electricity is the "lifeblood" (Huber 2013) of commerce and industry. Indeed, the relationship between oil and electricity was quite direct, since at the macro level Tanesco imported enormous quantities of diesel, heavy fuel oil, and jet A1 fuel to keep the IPTL, Symbion, and Aggreko plants running, and this would have large ramifications for the national debt, as I touch on below. As early as March, the Confederation of Tanzanian Industries reported that many factories were forced to completely shut down operations.[27] The price of commercially farmed sugar in northern Tanzania's Kiru valley increased 20 percent due to increased production costs and due to the fact that "the power crisis added to the problem of insecurity"; local communities torched several sugarcane fields in the area.[28] Students of VETA, Tanzania's network of technical colleges, had difficulty finding hands-on training for computers and other electric/electronic equipment.[29] Dar es Salaam's tourism rankings fell precipitously and the IMF lowered its forecast of economic growth.[30]

Smaller industries in Dar es Salaam suffered as well. Meat and medicines spoiled, while DVD vendors saw their sales plummet; no one, one proprietor explained to me, would buy a movie without having it tested on the store's television first. Sellers of *mitumba* (secondhand clothes) complained that customers shopping in dark corners of the Kariakoo market could not properly see the colors of the clothes they might purchase.[31] Streets where welders and other artisanal manufacturers were located sat unusually quiet, whereas downtown commercial storefronts could become a "primitive farm of purring and noxious gas-spewing generators."[32]

Domestically, the continuous shutdown and resumption of service also played havoc with electrical appliances. "Recently," one columnist wrote, "while watching one of those stupid soap operas whose ends are predictable, the power went off but when it came, my junk of a TV exploded like the [Gongo La Mboto] bomb!"[33] The reference was to an event in February, when munitions from the Tanzanian People's Defense Force barracks, located near the peri-urban neighborhood of Gongo La Mboto, detonated.[34] Artillery shells rained down on houses and schools nearby, killing twenty civilians and injuring 145 more. One of the official explanations was that the power cuts had shorted the air conditioners which kept the munitions storage at a proper temperature. On my very first afternoon in Dar es Salaam, in the car driving away from the airport, Ally pointed out the damaged roofs and pockmarked walls as we passed through Gongo L'Mboto. Ordinary channels of "cool, clean, quick current," what Durham Peters (2015, 126) calls "vestal fire," could become dangerously divested of their capacity to channel and control.

As the weeks and months dragged on, these straitened circumstances gave rise to a distinct ecology of adaptation, one with its own rhythms and phenomenology. Living in an apartment in the commercial/market neighborhood of Kariakoo, I became habituated to the sound of generators as soon as I woke up, and a tide of expectation would rise in the evening. Whenever the power did finally return, kids in the street would cheer, on behalf—it seemed—of everyone's shared relief. Phones could be charged, fans cooled skin, and people were psychologically refreshed to move through the next day's trials. If the power was late to return, my Tanzanian roommate and research assistant Thierry, a bit of a fashion plate, would decamp to a friend's local generator-powered bar to iron his clothes. The cuts, it was often suggested, tempted people (especially men) to frequent bars in general. They could avoid "sitting in the dark," but not the strain on their wallets, nor the inevitable hangovers.

Thus while the cuts had varying effects depending on people's positions and resources, there was also a sense that people were suffering them more or less *together*, and this could impart a rueful solidarity. One columnist relayed the following example:

> A few days ago I had an important meeting to plan for, so got home at 5 p.m. and as usual there was no power. I assumed that it would come back but by the time I was going to bed there still wasn't any sign of it. I obviously assumed that it would be there in the morning. But when I woke up, there was still no power!
>
> I entered the boardroom apprehensive of my wrinkled clothes assuming that I already lost my opportunity to make that first impression.
>
> But on entering the boardroom, I was not alone, the chairman had a wrinkled shirt and so were two other important people. What a relief! We all smiled and immediately started ranting about Tanesco and its woes on our lives.[35]

To be sure, not all were receptive to such ambient solidarities. One night I was walking with my friend Hamedi, eldest son of a wealthy Arab-Shirazi family, down the streets of the Kariakoo market area, dark but for a smattering of lighted windows across the faces of the *maghorofa* (multistory buildings), accompanied by the rumble of their generators. "There," he pointed with some satisfaction. "In the dark you can see who are the men and who are the boys." And yet as much as I envied the apartments for having power, I also knew that they were burning through enormous amounts of money and thus could only be swimming against the tide. Against Hamedi's appreciation for the "individual competitive liberalism" embodied in the private generator (Larkin 2008, 244), I was inclined to reverse the figure and ground; those few who managed to keep a light burning only offset the vast darkness of the rest of us.

watanzania waoga (timid tanzanians)

For some bloggers and writers, the quiescent way their fellow Tanzanians behaved in such circumstances—that is, the way they responded to the state's own misbehavior—was deeply revealing of the national character. In January 2011, for example, the *Citizen* published an anonymous editorial entitled "With Such Meek Citizens, Who Needs Riot Police?"[36] Earlier that month, Tanzania's security forces had violently dispersed a peaceful demonstration by supporters of the leading opposition party CHADEMA (*Chama cha Maendeleo na Demokrasia*) in the northern city of Arusha. Why, it asked with a touch of sarcasm, would the government find it necessary to deploy violent force against its own citizens? After all, Tanzanians are famously *waoga*—peaceful, almost timid—even in tough times. The new year arrived with inflation, crippling power cuts, higher tariffs, and a court ruling that Richmond/Dowans, despite having provided no actual power generation back in 2006, was entitled to "demand its pound of flesh" in the form of capacity payments (see chapter 1).[37] Compare this, it suggested, to Algeria, where higher food prices had "recently led to what could be considered an uprising." But "whereas Algerian rioters are angered by the increase in the price of foodstuffs, Tanzanians are not angered [by the power cuts] . . . what do Tanzanians do? Some buy generators to improvise, others buy *vibatari* [matches] and *mishumaa* [candles] while those who cannot afford any source of energy settle for a sigh of indignation: '*Yote tunamwachia Mungu*' [we leave it all to God]. In such a country people will just sulk and move on with their lives."

Other writers concurred. "It is no secret that what gives our corrupt leaders the arrogance to abuse Tanzanians is their seemingly limitless gentleness/timidity," one blogger wrote. "People pay taxes—on their salaries, in VAT on purchase of goods, and through other vectors—but the benefits of those taxes are more apparent in the size of our leaders' stomachs, the value of their cars and houses, and the adornments of their mistresses."[38] Tanzanians may complain on Twitter, he continues, but there is no "civilized" (*ustaarabu*) way to solve the outages; it cannot be given like a gift but must be demanded. A following entry repeated many of these themes and then ended by noting that "fortunately, our colleagues in Senegal show us how problems affecting the country are being dealt with by the public." The post then linked to a news article detailing how rioters in the capital city Dakar had poured into the streets to demand electricity and stormed the office of the national utility company, Senelec.[39]

What then might account for Tanzanians' seeming "timidity" in the face of power cuts and suffering more broadly? A clue comes from a Kenyan writer who, back in 2000, was complaining that his own country's power cuts had become "worse than [in] Tanzania: I remember visiting Arusha a while ago and couldn't breathe because every shop had its own little diesel generator on the street, pumping out black smoke. 'Why don't you guys have electricity, are there no rivers in Tanzania?' I'd ask. '*Hizo ni shida za Tanesco*' [These are the problems of Tanesco], they would shrug, with a certain sanctimonious, communist fatalism."[40]

The notion that Tanzania and Kenya form a contrast pair, one taking the sober, socialist path after independence and the other the wild capitalist path, is a perennial one in postcolonial East Africa. The examples range from the mundane to the existential. Kenyans, a friend once explained, are said to purchase shop goods by demanding *nipe* (give me), whereas Tanzanians offer the more restrained *naomba* (I beg for). Ally, who was imprisoned in the 1970s for selling contraband cigarettes, and bore no love lost for *ujamaa*, once cited what seems to be an apocryphal debate between Nyerere and first President of Kenya Jomo Kenyatta, where the latter proclaimed *ninaongoza watu, sio maiti*—I lead people, not corpses. More profoundly, it is also noted that while Kenya shed blood for independence, it was given to Tanzanians. Writer Nkwazi Mhango captured this basic contrast as follows:

> Shortly after our independence, there was a lampoon between Kenyans and Tanzanians. Six years after independence, we (Tanzanians) determinedly ushered in the Arusha Declaration of socialism. Equality, national owner-ship of means of production, self-reliance and total emancipation of Africa became our main and noble vector. Indeed, we did a tremendous job to emancipate many African countries especially in [the] South[ern] African region, thanks to the late Julius K. Nyerere.
>
> Kenya, on the other hand, embarked on western capitalism—under which equality of human beings is utopia. Kenyans called us a "man-eat-nothing" society while we referred to them as a "man-eat-man" society. In Swahili, a society of *nyang'au* or gibbons! I still remember one radio presenter asking his audience: "In which country does everybody own a car?" They used to reply: Tanzania, just because we wore tyre-shoes famously known as *katambuga*. We were a time capsule in the region for everybody to laugh at![41]

To understand why Tanzanians simply "sulk and move on with their lives" in the face of hardship, it helps to consider the ways African socialist states configured the relationship between time, development, and collectivity. By 1967, it was clear that all newly independent African nations had entered the world stage hobbled by over a half-century of colonial rule. Rapid develop-ment was required to cut off history at the pass (Donham 1999; McGovern 2012b); Nyerere insisted that "[Tanzanians] must run while they walk" (quoted in Grundy 2017). And yet Nyerere also warned against the temptations of shortcuts, of haphazardly accepting foreign capital, which by itself could only be a kind of simulacrum of development.

> We have put too much emphasis on industries. Just as we have said, "With-out money there can be no development, we also seem to say "Industries are the basis of development, without industries there is no development." This is true. The day when we have lots of money we shall be able to say

we are a developed country. We shall be able to say, "When we began our development plans we did not have enough money and this situation made it difficult for us to develop as fast as we wanted. Today we are developed and we have enough money." That is to say, our money has been brought by development. Similarly, the day we become industrialized, we shall be able to say we are developed. Development would have enabled us to have industries. The mistake we are making is to think that development begins with industries. (Nyerere 1967, 26)

Here Nyerere is outlining what McGovern (2017, 13) identifies as socialism's "orientation to the future," its insistence on "[imposing] an orderly narrative of social and historical progress that aligns past, present, and future." For Nyerere, accepting foreign loans or hosting private investors is essentially to act rashly, without thinking it through. These are poison gifts that will simply reproduce conditions of neocolonial dependence and inequality, leading the country back to the colonial position it started at. Rather, true historical progress only happens when the forces of wealth and industry are channeled through proper social form: a self-reliant national collective wherein the labors of citizens are harnessed to state management. Rather than lapse into the catch-as-catch-can scrum of capitalist competition, the party-state offered a pact in which all citizens would achieve development (*maendeleo*). But this pact had its own internally coherent temporal logic in which past injustice and future redemption were causally linked via state control in the present. Citizens had to trust the state that any hardships they might endure were not merely pointless exercises in suffering but a kind of meaningful sacrifice that would ultimately ensure egalitarian ends. As with any sort of form, one submits to certain "constraints on possibility" (Kohn 2013, 158) in order to enable others.

Hardship and sacrifice were thus central to socialist nation-building. Practically, this meant various kinds of high modernist "slate cleaning" (McGovern 2012b, 164): prohibitions on rural settlement patterns, "laziness," "backwards" cultural practices, and "traditional" forms of chiefly authority (Lal 2017; Schneider 2004; Scott 1998), along with strict policing of urban migration, fashion, and personal wealth (Ivaska 2011). In its place the *wananchi* (citizens) received the propulsive developmentalism of *ujamaa* and identification with the charismatic image of the Nyererean state as the good, if strict, Father. Nyerere lacked weight and wealth, but this lack was the very thing around which Tanzanian identity was structured: the "food" he fed to the nation was garnished by the language and reflected image of moral righteousness and the desire for a socialist future.[42]

Such sacrificial logics unfolded in dialogic opposition to other postcolonial paths. Fellow African socialist leader Sekou Touré once proclaimed: "We prefer freedom in poverty to opulence in slavery" (quoted in McGovern 2017, 38)—a comparison of his own newly independent Guinea to neighboring West African countries like Cote D'Ivoire that were quickly embracing capitalist integration with their French former colonizers. For Tanzanians, likewise, Kenya's capitalist "man-eat-man" politics may generate wealth, but it betokens a kind of naïve immediacy, an approach to reality that begins and ends with the actual distribution of material things (i.e., either one has the car, or doesn't; either a people are truly independent or not). The socialist logic by contrast begins with the future, with a collective social form or principle that will eventually be "filled in" by material forces. Joel Barkan (1994, 23) notes, "Tanzania's pursuit of socialism entailed a conscious decision by the country's leadership to sacrifice a measure of economic growth to achieve a measure of equity."

As I suggested in the introduction, these "slate-cleaning," future-oriented gestures—this coming into form—might be understood as a kind of phatic communication. Phatic communication is "empty" in the sense that it suspends referential content in order to inaugurate, sustain, or repair the conditions of possibility for that content to circulate. That emptiness is what makes it generative—a fact I came to appreciate while learning Kiswahili and its many salutations. When we answer a standard greeting like *mambo vipi* (how's it going) with an equally rehearsed *poa tu* (just fine), when we observe the formalities, we subordinate ourselves to the form of the relation, bringing ourselves into alignment so that we can subsequently share more substantive kinds of messages. This is a deeply infrastructural practice, concerned with the setup or maintenance of channels through which the stuff of collective life—energy, information, affect—is sent and received. Tanzanians were continuously and reflexively brought into alignment with the state and with each other, interpellated into a vast sociopolitical piping that would be progressively "filled in" with industries and money.

One of the pleasures of collectively assuming social form, moreover, is the way it looks to an external gaze, however imagined or implied. Ali Mazrui (1967) famously called it "Tanzaphilia": Nyerere's vision of *ujamaa* was an inspiration to leftists, sympathetic academics, and even a couple of Black Panther emigres, all convinced that the East African nation was on the right side and even leading edge of history. When I first came to the country as a student learning Kiswahili, I would impatiently dismiss the many soliloquies where I was welcomed to "Tanzania, the land of peace and unity" (*nchi ya amani na umoja*). I thought this was a platitude designed to appeal to or reassure an American visitor/tourist. It was only later that I could really hear what this meant—how deeply a sense of stability and national unity is woven into the identity of Tanzanians on the international stage. Once the home base of southern African liberation movements, Tanzania continues to host refugees, to house the seat of the International Criminal Court, and to position itself as an exemplar in a troubled region. Indeed, an early national Tanzania symbol is the *Mwenge wa Uhuru* (Freedom Torch), meant to inspire hope in Mozambique and South Africa prior to their liberation movements. This is an obvious and powerful metaphor for double-consciousness, for seeing yourself through the eyes of others, and it moves through the grand avenues and little corners of Tanzanian public life.[43]

The wry "Kenyan" rejoinder to all this heroism was that the commitment to outward social form may well come at the expense of inner substance. By the late 1970s, Tanzanians' sacrifices were not exactly bearing fruit, the sense of orderly progress stalling out as they navigated the external economic shocks and internal postcolonial burdens described in chapter 1. Dar's ailing public transport, for example, remained a visceral expression of historical lag. By 1983, during the Operation *Nguvu Kazi* (Hard Work) campaign that repatriated unproductive urbanites to rural areas (Kerner 2019), residents were angrily writing to newspapers that they were being targeted as lazy "idlers" when in fact they were merely waiting around for the overloaded and consistently delayed bus system (Brownell 2020, 100)! Taking the bus thus involved that quintessential mix of resentment and identification that marks the condition of "cultural intimacy" (Herzfeld 2005). On the one hand, access to urban mobility was severely curtailed. Any given resident was not likely to have a car (except for the tire treads on their feet), which we can take as a gloss on the capacity to move at will. On the other hand, any *other* resident was not likely to have a car either; the reduced access was rationed, routed through the municipal transport network. This ambivalence was well captured in what Emily Brownell (2020, 100) calls "liturgies of waiting," the public airing and sharing of daily grievances that drew people together in a sense of ironic, even rueful solidarity. Yes, Tanzanians were eating nothing, but they were eating nothing together.

Forty years later, I argue, a similar sort of rueful acquiescence could hold for power rationing. By the mid-2000s, the electric supply in Dar es Salaam had become just as important as urban transport in allowing the city to run, whether through the explosion of SMEs (small-and medium-sized enterprises) like salons and garages, or through the flood of communication technologies like televisions, phones, and computers. And like bad traffic, all residents have endured the endless grinding dysfunctions of that supply. However, because Tanesco's transmission and distribution system has remained public, its shortages have generally remained public as well. That is, it has generally faltered in such a way as to preserve the social form that the flow of electricity traces out, even as that flow was diminished in its material force. In a blog post on the crisis, January Makamba, parliamentary chair for the Committee of Energy and Minerals at the time, evocatively, if unintentionally, captured this condition: "Once upon a time," he wrote, "there were no power cuts. Yes, those who were connected to the Grid were relatively few, but had power most of the time. Today, those connected to the Grid do not get power most of the time."[44] In twenty-first-century Dar, more residents than ever are connected to the grid, but they are often connected through the paradoxical experience of *not* getting power.[45] Residents on the whole may have been deprived of electricity, but not relative to each other. Such common burdens did not invite protest or riot so much as another round in a long history of complaining and coping—a "sanctimonious communist fatalism," perhaps.

looking upstream

For the critical-minded writers and bloggers discussed above, however, collectively complaining and coping with Tanesco's rationing in ways that seemed to channel the socialist past amounted to a misrecognition of how much *had* changed in Tanzania. Decades of liberalization had created an entrepreneurial class of business and political elites exemplified in the "emergency power" contracts of IPTL and Richmond. The *Richmondi* scandal in particular nourished a minor but growing current of political opposition. In the 2010 presidential election, CHADEMA had secured nearly 30 percent of the vote, a remarkable jump from the previous, single-digit showings it had regularly produced in 1995, 2000, and 2005. The power crisis that unfolded in early 2011 only further underscored this loosening of CCM hegemony, at least potentially. After all, Tanzanians had learned—bitterly and belatedly—that the 2006 power crisis five years earlier was not only caused by drought or a generalized postcolonial underdevelopment, but by elite malfeasance and self-enrichment at the highest levels. State-managed load shedding, however collectively borne, certainly feels less tolerable if one experiences it with the knowledge that state agents themselves have contrived and profited from it.

In response, CCM and its supporters attempted to frame the narrative of the power crisis in ways that aligned with a nostalgic, Nyerere-centric approach to reckoning with the dissatisfactions of the liberalization era. In early July, for example, cabinet minister Samuel Sitta, sponsor of the parliamentary report on the *Richmondi* scandal (and rival to Lowassa), denounced the crisis as the work of "wealthy corrupt individuals," declaring: "we will not let this one pass unchallenged, and we promise you that we will do all it takes to restore the integrity this country was known for during the time of Mwalimu Nyerere."[46] The country, he observed, was blessed with abundant natural resources, thus it could only be "due to sheer greed that the nation is now in darkness"—a nod to the way criminally expensive EPP contracts had in effect foreclosed investment in expanded generation.

This was indeed a familiar rhetorical gambit. After Nyerere's death in 1999, CCM began to lay claim to his popular legacy of incorruptible morality (Fouéré 2014).[47] This was a rather sanctified idea of the man, grounded in memories of what Emma Hunter (2015, 87) calls "the Nyerere of the Arusha Declaration." By the mid-1960s, the party was suffering from popular anger at leaders' corruption and dissensus at the top, embodied in the suppressed army mutiny of 1964 and in Nyerere's charismatic rival Oscar Kambona (exiled in 1967). Against this backdrop, the Declaration drew moral lines in the sand, hardening the fuzzier, attitudinal *ujamaa* that appeared in Nyerere's writings from 1962 onwards. If 1967 was otherwise a year of curdling postcolonial pessimism, exemplified in the toppling of nationalist leaders like Ghana's Kwame Nkrumah, then the Arusha Declaration went a long way to consolidating moral-political hegemony by rejecting Western foreign influence, embracing nationalization, and excising unscrupulous/capitalist "enemies" (*maadui*) and "parasites" (*wanyonyaji*). As we saw in the last chapter, this was exactly the playbook that Kikwete leaned on to mitigate the scandals of emergency power: the cancellation of the NetGroup contract and rejection of Tanesco's privatization in 2006, the sacking of sundry Tanesco managers from 2007 onwards, and the forced resignation of his PM Edward Lowassa in 2008. It was this same playbook that Sitta drew upon in his address. The message was clear: after flirting with the pleasures and problems of liberalization, the party was returning to the righteous path, to the integrity of the Nyerere years.

Other "returns to Nyerere" coursed through public talk. A few weeks later, I attended a "Meeting to Debate the Electricity Problem in Tanzania" at Karimjee Hall, sponsored by The Institution of Engineers Tanzania. One presentation began:

We have a moral and ethical dilemma in Tanzania, which cannot be modeled in a computer or solved by a scientific formula. In the absence of the

Arusha Declaration, how do we go about solving this dilemma? One of the conclusions we reached is . . . that we need practical political will and patriotism to advance in tackling the power crisis.

Over and over again, it has been said that a nation shall never develop using foreign expertise. Recently there has been a conference organized in Rome on the Power Crisis in Africa. Another conference of the same nature is due in Uganda next month. All these are not for our benefit. They are looking for opportunities for their development. The majority of foreign companies (be of construction or consultancy) are there to market products and technology from their countries. While we have very good transformers from TANALEC, very good cables from Tanzania Cables, very good electric poles from Sao Hill to mention a few, foreign consultancy firms tend to recommend foreign products, and if you follow up, the products recommended are from their countries of origin or from sister companies and the like. Let the Patriotic Local Consulting firms be given the power of attorney to take the lead in power projects right from inception to commissioning and in a given spell of time the difference will be obvious. We envy the Koreans, Indians—they trusted their local experts: they made blunders, they were given opportunities to correct and they made it. In the same way, we can use experienced consultants to be the envy of others.[48]

Later that evening as I was sorting through pictures of the event, Hamedi happened to see a shot of the speaker, a plump, elderly gentleman perhaps in his sixties with a quasi-Nehru style jacket and laughed: "now that's what a real Tanzanian engineer looks like, you don't see those guys very much anymore." When I told Hamedi about the "patriotism" of his presentation—denouncing foreign companies, etc.—he nodded and explained how back in the 1960s and 1970s engineering was an immensely popular career choice. "You have to understand," he said, "those guys were around when the TANU kicked the British out, they wanted to build the country up."

A slightly subtler "nostalgia for the future" (Piot 2010) was evident in the explanations offered by January Makamba, the minister who had to my mind so felicitously described the crisis as a state of being connected to the lack of power.[49] Makamba confidently explained that the roots of the power crisis lay in the misguided attempt to privatize Tanesco. During the long period (1996–2006) of stalled privatization, he argued, no new state investment was made in the energy sector, during which time demand grew precipitously, leaving the sector permanently on its backfoot. As recounted in the last chapter, the story is considerably more complicated, given the imbrication of elite party politics with EPP and IPP contracts, but once again this telling has a certain narrative parsimony that attributes the power crisis to a naïve faith in Western-style privatization and, implicitly, to an abandonment of Tanzanian self-reliance.

These accounts of the causes of the crisis evince the same neo-Nyererean sensibility that animated much political discourse in the 2000s. As such, they politely sidestep the important role that a structurally decentralized, panic rent-seeking CCM itself played in producing that crisis. These tensions are occluded by invoking a unified Tanzania that, beset by outside forces, must preserve its integrity. As one editorial proclaimed: "Let us look at the power problem as a matter of life and death. We should tackle it with all our might, even if it means calling on Tanzanians to tighten their belts as they did four decades ago during the war against Idi Amin of Uganda."[50] Tanzanians, just as they were back in the 1960s and 1970s, should be in it together, united under the aegis of a ruling party and enduring a common affliction.

patronage vs. rights

Were they indeed in it together? The relatively quiescent response to the power crisis was a function of the complex field of signals that members of the public received from the grid, the state, and from each other. Urban residents certainly resented the hardships of power rationing, and some offered ambivalent commentaries on the national historical bargain they struck. And yet, for all that, the plodding normalcy with which we all accustomed to those hardships seemed to honor that bargain, and allowed the city to proceed as if the government were acting in good faith. Much of this, I've suggested, had to do with the manifestly collective quality of load shedding, buffered as it was by the sheer volume of "declarations" that accompanied it—the return to Nyerere, the perilous water levels at Mtera, the Stakhanovite invocations of hundreds more megawatts brought on by new and emergency power projects in the pipeline. This is not to say that these explanations were entirely convincing, exactly. After all, the *Richmondi* scandal had unfolded just a few years earlier and CCM held both legislative and executive power by the narrowest electoral margins since the advent of multiparty politics; it spoke from a position of diminished credibility, especially on matters electric. But insofar as these messages didn't outright contravene the "facts on the ground," they at least indicated the state was making some effort to court and communicate with the public. Tanzanians were suffering, and perhaps suffering in ways that were not entirely justified, but some combination of the state's reparative gestures, the reflexively collective experience of rationing, and their own socialist inheritance allowed them to retain, maybe in spite of themselves, a "sanctimonious communist fatalism," a sense that their leaders were still, however faintly, within the horizon of a collective project. And in any case, what was the alternative? Would it prove to be any better?

In her "ethnography of hunger" in rural Singida, central Tanzania, Kristin Phillips (2018) writes perceptively about how, over the 2000s, opposition politicians like Ibrahim Lipumba of the Muslim-dominated Civic United Front and Tundu Lissu of CHADEMA mobilized a "rights" (*haki*)-based idiom, rather than a "patronage" idiom more characteristic of CCM (and indeed of many postcolonial nationalist parties). In the logic of the latter, CCM is the father who feeds his citizen children who are expected, categorically, to remain loyal to the family. By contrast, a rights idiom tends to emerge when a population can no longer see "the state as father and the nation as family, but [rather sees] the state as parasite," whose exploitations must be countermanded through action and advocacy (Phillips 2018, 163).

In 2011, exercising one's rights certainly resonated with a regional and even global moment of popular uprising against parasitic, predatory powers (Branch and Mampilly 2015). Alongside the Arab Spring and Occupy Wall Street, walk-to-work protests were unfolding to Tanzania's north in Kampala, Uganda (Ojambo 2016), while carnivalesque occupations had overtaken Maputo, Mozambique to its south (Bertelsen 2016). Privileging action over sufferance, demand over supplication, these uprisings aspired to the ruptured time of the Event (Guyer 2007, 417), when situations miraculously "break with the past" (McGovern 2012a) and become fluid and open. The talk of Tanzanians handling the power crisis like the Senegalese or Algerians—that is, their rioting—has to be seen in the context of this historical moment. Whether an "African Spring" might arrive in Tanzania, and what might provoke it, was an ambient question.[51]

And yet as the jab at Kenyan politics as a "man-eat-man" society intimates, there are also traditions of political thought that appreciate the dangers of the breaking with the past and the ways such a break might spin out into social fragmentation. Consider McGovern's (2015) discussion of the historical imagination amongst his interlocutors in the forested region of Southwestern Guinea. The Loma villagers he interviewed divided up historical experience into periods of relative *booyema*, which translates to the "strength of one's own arm" and stands for personal liberty and striving, and *ziiεlei*, a "cool heart," which stands for collective peace and security.[52] The precolonial era of slave raiding and warlords in nineteenth-century West Africa tended toward a dangerous liberty that translated to upward mobility and enrichment for the few and rampant insecurity for the many. This Gilded Age of Afro-capitalism was not consigned to history but rather resurfaced in the post-1980s period of deregulated economics and government and, more specifically, in the resulting wars that enveloped the region in the late 1990s and early 2000s (Richard 1996; Ellis 1999). Guinea remained an exception in the region, in part, McGovern (2017) argues, for the way a shared socialist history, repressive though it was, preserved a habitus of collective interdependence.

Somewhat analogous dynamics can be found in postcolonial Tanzania. Nyerere's occasional references to the communalism of "traditional" African village life notwithstanding, Tanzania's socialist government had less in common with a precolonial politics characterized by the depredations and enrichments of the slave and ivory trade than with the postwar colonial state's ambitions to monopolize migration, trade, and taxation (Burton 2007; Brennan 2008). And yet taking the reins of the colonial state was a site of genuine ideological investment, even as it was resented for its incursions into people's autonomy and for the material deprivations it entailed. Autocrats like Idi Amin of Uganda or Jean-Bédel Bokassa of the Central African Republic Africanized the worst elements of colonial coercion, "styl[ing] themselves as 'chiefs of chiefs' using a backward looking utopian idiom." And yet as burdensome and coercive as socialist regimes could be, leaders like "Toure and Nyerere styled themselves as servants of the people's demands for justice and fairness, looking forward to a utopian future" (McGovern 2017, 223).

Thirty years of liberalization strained the plausibility of that utopian future. But to disregard the various explanations and exhortations and forgo the rueful solidarities of national crisis was a high-risk, high-reward gamble. Opposition—even simple electoral opposition—could escalate into a dynamic of power grabbing and repression that might look less like the collective effervescence of the Arab Spring and more like the repressive kleptocracies or roiling ethno-politics of neighboring postcolonies like Uganda, Kenya, or Democratic Republic of Congo. Yes, seized by frustration and revolutionary spirit one could act "in the now," but what might be lost the day after?

The columnist Mlagiri Kopoka circled this tension in an August 26, 2011, column titled "Did the West Tell All about UK Riots?" The title refers to the riots of August 6–11, 2011, which saw looting and destruction in London and other cities after the police shot and killed a local man of Afro-Caribbean descent named Mark Duggan during an arrest. It was like, Kopoka muses, "the Arab revolution had finally reached Europe." Yet Kopoka finds his desire to follow global current events is undercut by his own country's disconnection. Load shedding meant that he was not up to date on the international news that day. Luckily, there is a small TV at the front of a *daladala* he boards heading home from work. At first there are just a few passengers so he asks the driver, "as if the other passengers had selected [him] to represent them," to turn down the loud *bongo-flava* music coming from the speakers so they could hear the news, and the driver obliges. As more passengers pile in, tiffs ensue. A seated man asks a young woman unlucky enough to be standing in the aisle if she could "bend a little" so he could see the screen, to which she retorts that if he wants to see the screen so badly he should just switch with her. Other passengers joined in to agree with the woman, mocking the man and the sexual undertones of his request that she "bend." As the passengers jostle, Kopoka's thoughts

> switched back to the rioting youth and what was shocking was the fact that the mobs in Britain did not have a civic agenda. They were out on the streets to simply cause mayhem by smashing, looting and burning property. Then I asked myself whether the western media was hiding some of the truth about these riots because they were taking place in their own back yard? I wondered if the same type of riots were taking place in a third world country like Bongo the stories would have been the same. Suddenly I lost my view when a certain woman entered and stood right in front of me as a feminine voice said, " *ii jamani pole* [hey easy!]. You are hurting me!"

This is a suggestive scene with many interwoven threads. The first is the quintessential experience of riding a *daladala* in which passengers are stuffed cheek to jowl and, against all odds, always seem capable of adding one more person. Like many African metropolises, Dar es Salaam's public transportation has grown both better and worse since the 1970s. For residents who live out in the city's rapidly expanding peri-urban areas, evening and morning commutes on the cities' now plentiful, privately operated buses are still jam-packed and can last two to three hours; finding ways to stand in such cramped circum-stances can be agony. I had often thought jokingly to myself that *ujamaa* was alive and well in Tanzania, and that one simply needed to board a *daladala* to experience it. Sometimes after a drink or two of Safari lager I would try the line out on friends, usually to bemused (or perhaps just charitable) chuckles. And yet I was not alone in thinking along these lines. In his "Essay on the Share" (2017, 25–26), Ferguson reflects on the African minibus taxi as a site of "a kind of shared sociality, where certain minimum standards of civil conduct are almost always respected [based on] on a kind of accidental co-presence . . . this adjacency imposes a non-trivial sociality that entails real obligations and a more or less continuous set of pragmatic adjustments, [given that passengers must] yield a precious share of that scarce, tightly packed space. . . . [W]e must make ourselves less comfortable, simply because someone (with the same needs as we have) has appeared."

This allocation of space is not exactly planned, nor is it particularly comfortable. It is, as Ferguson goes on to observe, defined above all by irritation and complaint. But then what can one do? One is obliged. In this sense, the random movements and adjustments are not mere physical hardships but meaningful sacrifices, pragmatic expressions of a governing logic, a commitment to sociality as such.

We might, then, take Kopoka's experience of crowding around an international news broadcast on a cramped bus as an image of Tanzania's social contract in general. Frayed and meagerly provisioned as it is, people are still making pragmatic adjustments that signal and preserve its form. Brits, by contrast, are not riding the bus of state but, as it were, torching it. The "mayhem" that has unfolded across their cities—and the police killing that caused it—signals no such pretense of collective obligation, such that all that remains are the risks and rewards of aggressive force.

Of course, commitment to form can only hold for so long. Alongside international news and *bongo-flava* hits of the day, the city's radio and television stations also broadcast regular *matangazo* that 2011 was the fiftieth anniversary of Tanzanian independence. For some, these declarations could only ring hollow: "this nostalgia about turning 50 is being shoved down our throats. . . . Is this constant attempt at reminding us that we are turning 50 reflect a fear that we might forget? Is the fear that regular people aren't quite demonstrating their excitement at Tanzania's semi-centennial?"[53] The celebrations of Tanzania threatened to lapse into a notional promise that would never be substantiated, a wire without current, "a mere sequence of sounds or written shapes without [any] quality of life that animates it."[54] Or perhaps, most diabolically, those words could be cover for their opposite. Perhaps the parasites that exploited the nation were not external to the government still committed to a socialist morality, but rather one and the same. As Mhango writes:

The Kenyans we used to laugh at at least have something to show for their paradigm whereas Tanzanians have nothing to show but regret after some few devils miserably vended every mineral, animals, logs, banks, parastatals and what not for their own selfish interests! The enemy who tells you; "I am your sworn enemy" is better than the one who pretends to be your hubris [*sic*]. Ours have sedated us with sweet words—so as to [have us] mistakenly regard them as our saviors while they are our slayers![55]

Tanzanians who followed out this line of reasoning emphasized the necessity to take the political process into their own hands, whether via demonstrations or involvement in opposition politics. But in ways both conscious and subconscious, I think, it was also recognized that to do something about this in any dramatic way would be to risk the collective commitment to peace. Against this background, even "sweet words" and a lingering socialist habitus had a certain rueful value. Why didn't Tanzanians riot? Perhaps, at least for the moment still, it was better to eat nothing, together.

are we rationing?

In February of 2012, Kopoka penned another column called "Bongo: Land of Broken Promises."[56] It begins with an ode to the tranquility of his household over the past few months. After the power blues of July to October, Parliament passed its emergency power plan and now, mercifully, Kopoka narrates, the electricity supply has stabilized. His wife has even resumed her defunct ice-cream business. But recently, things have taken a darker turn. The lights seem "possessed by some evil spirits judging by the way the power keeps flashing on and off as if it is controlled by some demonic powers." The cuts keep interrupting a televised address by the Prime Minister regarding an ongoing doctor strike, and in frustration his wife groans that Tanesco should just keep its power "instead of playing these on and off games"—the surges had probably broken the refrigerator. Kopoka writes: "I opened the fridge door and looked at food and drinks that were beginning to decompose. I felt the anger in me reaching boiling point. I looked at the TV; the PM was trying to elaborate a point outlawing the doctors' strike when the lights went off yet again. In frustration I took an egg from fridge and threw it hard across the room. 'Pow!' In the darkness I heard it smash the TV screen!"

If one thinks of the distribution of power outages as a kind of national EKG, then Dar's generally remained within a reasonable bandwidth for late 2011 and early 2012. But soon thereafter, electric current was once again refusing to stay submerged in the background of everyday life: once again, appliances were blowing out and food was spoiling. Only Tanesco had not declared any rationing, and instead of long dips of inertia, the resurgent cuts were mysterious staccato flickers.

Thus we have the setup for a clever commentary on this predicament, which hinges on the presence of the prime minister, talking about the seemingly unrelated issue of the doctors' strike. The PM's voice and the power cuts seem to subtextually color, even ventriloquize, each other. The cuts are a kind of puncture, deflating the paternal aura of the PM's speech. But for Kopoka, they persist and even grow in the imagination in a way that suggests more than misfortune or mistake. Where the PM falls silent, the electricity supply *itself* seems to take up speech, flickering so wildly that one might well divine some hidden intelligence behind it. No wonder his wife says she rather prefers having no power at all. How can she believe that the state is acting in good faith, when the power cuts seem like a "game" they are playing? Ultimately, we are left wondering if the cuts are the ominous unspoken message of the prime minister's address: the government is not in the business of keeping promises. Indeed, the doctors' strike was shut down soon after, not through patient government debate and explanation, but by the brutal beating and hospitalization of one of its leading organizers, Dr. Steven Ulimboka.[57]

During my fieldwork in Dar in 2012 and on through 2015, and amidst these occasional acts of repression, power interruptions waxed and waned without ever metastasizing into full-blown crisis. A series of public surveys conducted by the NGO Twaweza gives a sense of the phenomenon. In February of 2012, 50 percent of respondents reported power interruptions on four or more days during the prior week. Twenty-two percent reported an outage every day of the week, with over half of those outages averaging six hours or longer—this up from a survey conducted in 2010, in which only 25 percent of respondents reported outages lasting longer than six hours.[58] In short, we might hazard that minimally, power outages could be expected at least twice a week for two hours and probably closer to three times a week for four to six hours. At the extreme end, one resident reported that his neighborhood could "be sure that within 24 hours [it would be] likely to get electricity for eight hours spread over one day and night. There are frequent cuts within 24 hours."[59]

This formless no-man's land between smooth(ish) functionality and regimented emergency measures was in some respects attributable to the fiscal aftermath of Tanesco's emergency power plans. To pay for Symbion and Aggreko's 200 MW of additional power, not to mention accumulated arrears from prior IPP and EPPs going back to the 1990s, Tanesco was requesting a state-backed loan of Tsh. 408 billion (~$272 million). Until then, the cash-strapped utility had little money to invest in network maintenance and repair, leading to a surge in network breakdown. Hence for much of this period, Tanesco tended to emphasize that the "outages were caused by technical problems [blown transformers, defaulting cables, generator breakdown] and not rationing."[60] Occasionally, Tanesco *did* declare a brief period of "rationing" as in March 2012 when Aggreko temporarily withdrew 100 MW from the grid due to nonpayment, or in November 2013 when Songas shut down its gas wells for maintenance.[61]

Whether deemed rationing or not, what all of these mounting interruptions had in common was their unpredictability. One blogger's description of the Songas "maintenance period" sums up the experience:

Last week, Tanesco announced a 10-day period of power rationing. The rationing was due to maintenance taking place at the Songo gas plant, which generates electricity for the national grid. The notice was sort of a disclaimer to say "we're going to cut it out at any time we see fit, and you'll have to deal with that." So in those 10 days, we never really knew when the power would go out. Sometimes it went out for an hour or two. Other times it went out for entire days and nights.[62]

And yet this was not that different from "ordinary" load management. The combination of disrepair, temporary supply deficits, and occasional service interruptions for maintenance added up to a kind of continuous peppering of outages, sometimes approaching intelligibility (e.g., "Tanesco has announced a maintenance period so we will have to expect some kind of interruption"), and sometimes falling back into something more murky ("the power has been out for an hour or two and then a day or night, but Tanesco has not said anything one way or the other"). Any explanations were vague and after the fact—usually accompanied by a conspicuous insistence that what was happening was *not* rationing qua systemic deficit/crisis of generation.

And it was the not knowing that frustrated people. As one interlocutor grumbled to me, "we just want *matangazo*, that's all. 'Tomorrow, starting at this time, the power will be out for repairs.' But if you find if you're dependent on business [it ruins you]. You ask yourself if there's no *mgao* why the silence? Here we're not like you [Americans/Europeans] . . . I don't know why it is that our leaders don't lead." This woman was one of the few in her area with a connection to the national grid. Along with selling cold drinks from a large refrigerator, she and her husband had invested in an electric pump to sell water to much of the neighborhood. She was especially focused on the logistical practicalities of power loss. But her follow-up musings—why the silence? What does it say about our leaders?—reveal the way the power supply indexes the overarching political compact between citizen and state. Provisions will wax and wane, but preserving a horizon of trust against which that waxing and waning becomes intelligible should not.

In that horizon's absence, the outages taxed the public nervous system. Will the problem in the supply resolve itself? Drag on? Metastasize into a full-blown rationing like the year before? People cultivated practices of infra-divination (Trovalla and Trovalla 2015), an "everyday hermeneutics" (Larkin 2018, 193) that tried to make sense of things, often comparing notes across neighborhoods. As one resident in the Tungi neighborhood of Kigamboni ward described to me:

> Here in Tungi you hear the fridge pop, and it cuts out. So you go ask some-one nearby if his or her power is out. "Yes" [he'll say]. It's truly out every-where. So you wait a little while . . . and then call someone up in Magomeni or Kariakoo, "Hey what's going on with the power?" "You hear those gen-erators? There's no power in the whole city." . . . We say that here in Tungi we are oppressed, but yesterday I went to my mother's in Magomeni [and was surprised]. She said for the last three days "it's been just as you see it now: pure heat, and no power here."

An unexpected power cut arrives as a series of sensory impressions: the whir of the fan stops, the television cuts short, the street grows quiet. The meaning of these sensations is not always clear. Living in an apartment in Kariakoo during most of 2012, I found myself developing a kind of electrical habitus. If the power would go out once or twice a week for a few hours, it usually did not bear following up. I might peek across the hall or out the window to check that my neighbors' lights were out or that their TV had gone silent, thereby confirming the problem wasn't confined to my place alone. In other cases, the texture was unusual enough to warrant inquiry. The power might go out for two days, at an inconvenient time of day at night, or in rapid daily succession. I might play detective, noting the length and time of the last outage, or correlating it with recent pronouncements in the news. Just as my interlocutor from Tungi described, after a little while one begins to get antsy. How much longer will the power stay out? What is happening anyway? In this way, the threshold of accommodation for power loss is not (just) related to its intensity, but its degrees of intelligibility. Short, relatively infrequent cuts "go without saying," and for most of daily life that is exactly what happens. But as they begin to accumulate, their effects—lost income, rearranged schedules, boredom, anger—become too much. People demand that misfortune identify itself. Or, to use Povinelli's (2011a, 132) formulation, they demand that a "quasievent" be registered as a full event. They will likely ask each other that familiar question: *kuna mgao*? Are we rationing?

For some, the absence of an answer to that question in 2012 signaled that we undoubtedly were—though this perforce made it a different *kind* of ration-ing. Terms like *mgao bubu* ("mute" or covert rationing) and *mgao feki* (fake or contrived rationing) began to circulate, with the two meanings often intertwining. Speculations about collusion with power generation companies or candle importers abounded, with varying degrees of seriousness. Writer Nimi Mweta claimed that the increase of random power cuts and the refusal to announce/implement rationing was a kind of subtle "pressure method" on the government, designed to ensure that subsidies would continue to flow to Tanesco to pay for expensive emergency power.[63] This interpretation was given credence when the MEM publicly accused Tanesco management and allies in Parliament of conspiring to artificially create shortages and thus profit from kickbacks on lucrative tenders for diesel and jet fuel (see chapter 1).[64]

These revelations resonated with other evidence for foul play, namely that wealthier neighborhoods received discernibly fewer interruptions (see figure 2.1). Even during the power crisis in 2011, a Tanesco spokesperson had to assure the public that despite angry grumbling about appearances to the contrary, no residential or commercial areas of power were getting more than others. If it seemed otherwise, it was only because those areas were in or adjacent to "nationally sensitive areas" such as state ministries.[65] A year later, when I went to visit a friend at his house in the popular Swahili settlement of Vigunguti, he pointed to a wealthier area nestled up in the hills above him and explained how he could always see their lights on when his own street was dark. For residents living in areas like Tungi and Vigunguti, it was common sense that they were "oppressed" by the power supply. In August 2012, a resident named Samwel Munro (@murosam) tweeted: "Who says there's no rationing? Tanesco? I'm living testimony that there is! Or maybe we should say that in the area I'm living in there's fake rationing."[66] Samwel senses some illicit plan to withhold power, perhaps even just for his area alone—why else, he asks, would Tanesco keep silent about it?

In the postsocialist metropolis of Dar es Salaam, electricity has become a basic commodity of everyday life and thus subject to a popular moral economy; its regular flow is both the sign and substance of the state's ongoing commitment to its citizens. Scheduled load shedding at least preserves the form of that commitment, signifying a basic respect for and recognition of the public. Conversely, after a certain point, a run of outages that authorities evidently cannot or will not explain is not (just) a material deprivation but a social insult.

In his reflections on the Levi-Straussian theme of "living with others," Marcel Hénaff cites a passage from Wittgenstein's *Philosophical Investigations* that helps drive home the difference: "Why can't my right hand give my left hand money?—My right hand can put it into my left hand. My right hand can write a deed of gift and my hand a receipt.—But the further practical conse-

FIGURE 2.1 A Tanesco worker redirects current from "Walalahoi Street" (neighborhoods of the poor and dead-tired) to "Vigogo [Oligarch] Street"—wealthy neighborhoods like Oyster Bay and Mesaki.

quences would not be those of a gift" (quoted in Hénaff 2013, 76). Giving, in other words, "is only possible if one assumes that giver and receiver are not the same person. I cannot be the beneficiary of a gift (or more generally of a prestation) that I give to myself" (Hénaff 2013, 76). Outages turn that logic inside out and upside down. Elites *can* "gift themselves" by shuffling public money from the state's right hand to its left (via fuel tenders, budgetary allocations, and emergency power contracts) rather than reciprocating it back outwards, as electricity. In turn, residents—especially poorer residents—are no longer autonomous persons to be respected but extensions of elite desires, reduced to conduits of that self-giving, to mere props or playthings.

The idea that Tanesco considered residents mere objects to be (mis)used and abused found vibrant expression in popular culture. In late February of 2013, for example, "Tanesco" opened a Twitter account[67] and at first many mistook it for the official one, inquiring as to why their power was out. To their surprise, the answers were vindictive, glowering, and often hilarious. One of its tweets informed customers that Tanesco was cutting the power as an exercise to help citizens prepare for the "heat of the grave" in death (*joto la kaburi*).[68] Another observed that "Many of you who are complaining about our power cuts have just come to the city recently. Have you forgotten how you were living all those years in the village?"[69] The interactions with other users were in the same vein. When a woman tweeted at the account, thankful that her power had returned, it replied that she better hurry up with the iron because they would be soon shutting it off again.[70] Bewildered, one user wrote: "the idiotic way you answer us . . . I've never seen such a thing," to which the account snapped back "when you ask idiotic questions what do you expect?"[71]

If a power outage could speak, what would it say? Like the "evil spirits" Kopoka senses in his flickering lights, we see here (and on Twitter no less) a tongue-in-cheek rendering of what it feels like to be subject to a swirling force without form. Each unexplained cut pecks at the sense of basic respect and recognition residents feel they deserve from their leaders, keeping them off-kilter and reproachful. Taken as a whole they intimate a simmering, antagonistic energy. It is as if those forces, by virtue of the very fact that they don't "add up to anything," were a single omnipresent agent in whose eyes you are always already guilty.

This swirl of formless power cuts might be located in a longer lineage of demoniacal actor-networks like rhizomes, swarms, crowds, and contagions. Across philosophy and myth, media theorist Alexander Galloway observes, such forces "bring punishment, but not the kind of retribution wrought by . . . 'modern' juridical power. . . . They bring only the punishment of the ages" (2014, 57). He cites the Furies, those "gods of the incontinence of form," who were depicted by Aeschylus as a "bloody ravening pack" stalking their victims (Galloway 2014, 67), and Nietzsche's den of fanged tarantulas, who, Deleuze observed, "[preach] equality (that everyone become like [them]!)" (quoted in Galloway 2014, 58).[72] Such figures are quintessentially anti-hermeneutic; the search for any intelligible pattern or justice within them is always met with the same implacable answer: no. As McKenzie Wark (2014, 156) says of the Furies: "they can't be communicated with. They do it to us, not with us." As I show below, the experience of being subjected to electricity's incontinence of form, its punishing infliction of contingency, would come to be in keeping with other acts of a silent and silencing state.

political disconnect

As Kikwete's second presidential term (2010–2015) moved toward its conclusion, the dynamics of unpredictable power outages could turn from merely frazzling to ominous. The use of punishing force against oppositional voices, from striking doctors to peaceful rallies, seemed to be increasing; power interruptions were an echo and medium of such violence, as well as in at least one case its justification.

On Friday morning in October 2012, for example, around 10:15 a.m., the power dropped out along my street in Kariakoo. Approximately two hours later, as I wrote in my fieldnotes, "I was about to walk down the steps [of my apartment building] but something in the air didn't feel right." A minute or two later, I heard a deafening series of booms and, adrenaline racing, peered out the window to see hundreds of people flooding down the streets. As I would later learn, riot police had stationed themselves outside the central mosque a few blocks over. There had been growing anger at the arrest of "hardline Islamic cleric" Sheikh Issa Ponda the week before,[73] and apparently an anonymous ultimatum had been sent to the police that if he were not released, "protestors would not be responsible for what happened after Friday prayers."[74] After a heated confrontation with the exiting worshippers, police fired tear gas canisters into the crowd, and the next hour or two devolved into smashed windows, fires, and the violent rounding up of protestors. I couldn't help but wonder: was the power cut off in anticipation of the confrontation? Power went out all the time, of course. Was it a coincidence it went out just then?[75]

A few weeks after this episode, Kikwete announced the construction of a massive new liquid natural gas processing facility and power plant in Dar es Salaam, fed by a pipeline connected to the gas lands of the southern region of Mtwara (see map 1.1).[76] Since the mid-2000s, the fossil fuel industry's growing interest in Tanzania's vast energy reserves had stoked hopes for a miraculous new era of national wealth and, more specifically, that the country might finally get off the treadmill of emergency power and resolve its generation supply. Yet the pipeline was a surprise to residents of Mtwara. Since 2009 the state had promised that a natural gas processing complex and 300 MW generation facility would be sited in their region as part of a broader effort to stimulate the historically underdeveloped economy. Massive demonstrations against this broken promise flared in December 2012 and January 2013. In May 2013, "minutes before the Minister read his budget [confirming the pipeline]," radio and television signals went dead . . . residents with satellite TV, unaffected by the blackout, quickly spread the news via text messages. Riots broke out" (Mampilly 2013). Over the next two days of civil unrest, demonstrators torched government offices and targeted homes of party officials, prompting the state to send in security forces and eventually the army. Hundreds were arrested and at least seven killed in the tamp down, including a pregnant woman (ILPI 2014).

In one sense this was a classic case of distributive politics. In a speech haranguing the protesters, President Kikwete declared: "Natural resources, regardless of the region where they are found, are the property of all Tanzanians" (cited in Mampilly 2013). By contrast, Mtwarans resented the irony that their region's gas would be allocated to produce electricity that would never reach their underserved villages and towns. At the same time, they were not simply responding to raw deprivation. After all, people in Mtwara had long felt themselves to be historically marginalized (Ahearne and Childs 2018; Must 2018). The region was the site of an ill-fated groundnut scheme in the colonial era, and a central theater of villagization campaigns some thirty years later, coinciding with relocation efforts to protect civilians from the civil war in Mozambique just to the south (Lal 2017). And yet, as Zacharia Mampilly (2013) suggests, "for all the disastrous effects of the Mozambican war and forced villagization, people in Mtwara never abandoned the ideal of the Tanzanian nation. Under Nyerere, people largely accepted the moral logic of national sacrifice despite the unequal costs borne by the region. Indeed, their patriotic ardor remains far more substantial when compared with folks who live in Dar es Salaam. But the debate over the pipeline is fraying the national fabric."

The difference is that one kind of deprivation was accompanied by an ideological address, both overtly and in the very structure of the settlements themselves, which by virtue of their collective egalitarian nature sought to transform villagers from this locale or that into national citizens. The pipeline, by contrast, was a deprivation accompanied by resounding silence—indeed literally, during the media blackout, and also more generally. As one researcher who conducted interviews in Mtwara after the riots reflected: "It is important to note that in retrospect, most interviewees said that they would not have become so angry if they had only been given information and education on the rationale for the pipeline decision at the same time as it was taken. To them, this decision was tantamount to no local benefits and development, and at least a part of their anger was linked to a feeling of not being consulted or informed" (Must 2018, 99).

The greatest evidence that the pipeline was unjust was the state's refusal to explain it—to situate it within the temporal horizon of a collective future. As I have suggested, power cuts happening concurrently in Dar es Salaam played out this same dynamic, if in a much less concentrated way. Instead of one citywide blackout, the mercurial presence of "fake" and "silent" rationing created a kind of flickering atmosphere, an aura of active exploitation and disregard. Over time, these unpredictable interruptions threatened to reduce urban residents to mere objects to whom things are done, rather than subjects with whom the state communicates.

conclusion

In 2013, a joke was making the rounds through the SMS networks, blogs, and Facebook pages of Dar es Salaam. It went: "It's us who know all about Dar life! You all out there in the village, you slaughter the chicken and it's eaten in the same house. Here in Dar, chicken is slaughtered in Kisutu, the neck is eaten in Tandale, the wing is eaten at Tabata, the thigh at Magomeni, the gizzard at Kigogo, the legs at Mbagala, the ribs at Manayamala, the intestines at Kimara, and the head at Gongo la Mboto. Don't play around with Dar, our chow is like electricity: every person gets some!"

When I texted Ally asking what he thought of this joke he responded: "It's true. Oftentimes in the village if a guest comes and there is a chicken to spare, they'll eat it. If there's no chicken, they won't. They'll just eat greens (*mboga wa majani*) until perhaps a chicken is sick or dies." In other words, there is a logical consistency to the way food is distributed: people either eat (when there is food) or don't (when there isn't). In Dar, people are always eating—"everyone gets some"—but only in a shallow, literal sense. Some people *always* get the lion's share—those in neighborhoods like Magomeni—while others will always be consigned to survive on the odds and ends, the guts and heads. In other words, the difference between hunger and satiety is folded into the act of eating itself. Some people eat but never get full; others are never hungry but go on eating anyway.

In Dar es Salaam, residents experienced two broad modes of "eating" power cuts. Outages in 2006 and 2011 were marked by a high degree of suffering and crisis in which there was virtually no power; the city was, in effect, "villagized." After 2011, the government made strenuous rhetorical efforts to insist that there had been no rationing and in a narrow sense this was true enough—there was more absolute supply, and the city never descended into full-bore crisis. And yet what took its place was not consistent power but a series of longer and more frequent interruptions, ones clustering in poorer neighborhoods at that. In the short term, the latter was perhaps preferable (though recall Kopoka's wife, who judged that no power at all was better than such "on and off games"), but it nevertheless fed a growing mood of suspicion that the government was not just failing its citizens and pledging (however unconvincingly) to do better, but actively exploiting them. As with popular moral economies of bread or petrol, the manner in which electricity failed to arrive, the form(lessness) of that absence, became an important diagnostic of the social contract, and a crucial factor in shaping the moods and modes of response.

A few years later, in the runup to the October 2015 election, these two modes would converge in a kind of postsocialist crescendo. Kikwete had served the constitutionally mandated two terms, leading the way for a new CCM incumbent. Former Prime Minister Edward Lowassa campaigned mightily for CCM's nomination. Despite the infamy wrought by the *Richmondi* scandal, he remained a major power broker within the party, having built up an extensive *mtandao* (network) of patronage with churches and CCM's youth wing. But CCM's inner circle, headed by Lowassa's former friend, President Kikwete, tapped John Magufuli, a relatively unknown minister of public works. Magufuli's nickname was the Bulldozer, in reference to his reputation as an effective if pugnacious public servant reputed to be clean of corruption. Then, in another surprising turn of events, Lowassa defected to the opposition party CHADEMA, which promptly granted him candidacy.

While I discuss the ensuing Magufuli era in this book's conclusion, what is relevant here is the way that this dramatic election played out against a backdrop of power loss. Beginning early September 2015, Tanesco announced a week-long period of rationing throughout the country. Natural gas piped into Dar es Salaam's Ubungo generation plant, usually fed from the Songosongo gas fields, would be shut off and instead the plant would be fed from the newly completed Mtwara pipeline as a test run. But, as writer Elsie Eyakuze described at the time, the test week came and went and yet: the "rationing that was officially-not-rationing only got worse." Frustrated and annoyed, residents "encouraged [the utility] to just confess" what it was up to.[77] In early October, Tanesco announced that reluctantly it had to introduce load shedding due to low water levels after a weak rainy season the previous spring, and as such would be shutting down its entire hydropower network.[78]

Long days of load shedding ensued, as food spoiled in warming refrigerators, water pumps shut down, and clothes remained wrinkled and dirty.[79] Kikwete expressed "disappointment [at Tanesco] for his having assumed the presidency in [2005] when there was power rationing and leaving the office when the situation is the same."[80] Lowassa blithely announced that if elected he would discipline "lazy" Tanesco management,[81] a claim somewhat deflated by the power cuts that literally silenced him during some of his rallies.[82] In what would come to be his trademark, Magufuli leaned on a rhetoric of restorative discipline, promising to *tumbua majipu*, or "lance the boils" of corruption in the utility and beyond. Once again, power loss was both a failure and a promissory note.

But also, perhaps, a tactic. In early 2015, the Kikwete administration began its final year in power by passing a series of what we might call "blackout laws": the Cybercrimes Act and Statistics Act turned the publication of data or opinions not approved by the state into a criminal offense; almost immediately, writers and social media posters online were charged with insulting the president and were prosecuted (Parks and Thompson 2020, 4292). As the election campaigns ramped up, opposition supporters routinely complained about intimidation and the forceful dispersal of rallies. Against this backdrop, many urban residents found it notable that load shedding in effect cut off media coverage of the elections.[83] Eyakuze wryly captured the ways yet another round of collective power rationing could feel depressingly familiar to a country of "former socialists" but also, given the authoritarian turn of CCM in the past few years, unsettlingly different.

> Has there been "secret" power rationing in Tanzania for the past month and a half? Hey, now, careful. This is a sensitive topic for a highly strung, suspicious society. . . . Mtera Dam has been drying up and with it so has our capacity to produce power. Every year it is the same thing. Except when it is election year. During an election year, power rationing takes on an ominous tone deeper than the silence of households that can't switch on a fan during the hottest nights of the year. Rumblings of discontent started because decent, peaceable, TV-watching citizens started being deprived of political rally coverage. It's one thing to deny folks decent access to good health services and quality education. But when you start interfering with their ability to watch the evening news and recharge their smartphones? That's just asking for it. . . .[84]

As power cuts increase past a normal threshold of frequency and duration, residents at least expect them to resolve into some sort of evident and explainable pattern. But that resolution is sometimes delayed or never comes at all. Residents are left suspended in a strange liminal zone in which the outages are not (yet?) officially acknowledged or convincingly explained, stoking their agitation and suspicion. Power loss in the runup to the October 2015 election both made sense and didn't: while Tanesco had publicly announced rationing due to drought, its exercise still felt suspicious in ways usually reserved for the kind of "rationing that was officially-not-rationing" that had transpired the month before. Maybe the cause really was drought, but then again maybe the state would have found an excuse to "blackout" the power before the election no matter what.

This episode also highlights the recursive relationship between electricity and communication. Electricity powers the kinds of mass-mediated networks that now hold Dar es Salaam together in a state of collective experience. It is what is required to charge your phone or watch the evening news. Hence, to reduce the power supply during that election month was to be "kept in the dark" in more ways than one. It was to reduce the very infrastructure required to make sense of the state's behavior. In this sense, like the media blackouts that preceded the Mtwara riots, it marked a limit case of the postsocialist social contract, insofar as one of the hallmarks of socialism was the phatic experience of being in communication with the state and with each other, of contextualizing hardships in ways that held open a collective future. Thus if it is one thing to lack vital flows like electricity (or access to education or healthcare), it is another to endure that lack's lack of sense. For much of the 2005–2015 period in Dar es Salaam, the electricity supply flickered back and forth across that distinction, wavering between a hollow "form without force" and a capricious "force without form."

That flickering will set the scene for the chapters that follow. The second half of this book continues to examine the way the current-currency circuit mediates the social contract between citizen and state, but shifts down a level. We leave the headwaters of the national supply, with its fraught modes of generating and transmitting current, and descend to the street level of households, utility workers and electricians, and their modes of billing, payment, repair, and extension.

Cities with a long history may be called "deep" or "thick" cities in the sense that they are the historical product of a vast number of people from all stations (including officialdom) who are long gone. It is possible, of course, to build a new city or a new village, but it will be a "thin" or "shallow" city, and its residents will have to begin (perhaps from known repertoires) to make it work in spite of the rules. —JAMES C. SCOTT, *SEEING LIKE A STATE*

# of meters and modals 3

PATROLLING THE GRID

"Ah, a wedding. They're starting exactly at 7:00, just like proper Chagga people." It is Saturday night in an open-air bar in Mbezi Beach, a northern suburb favored by upper-middle-class professionals. I sit across from Mr. Njola at a set of plastic tables and chairs atop the concrete floor. Away from the denser core areas, the breeze is cool and pleasant. Soon the table will be crowded with beer bottles and a tray of roasted lamb flavored with lemon and salt. Friends and fellow regulars—officers, engineers, and professors—will stop by to say hello and share the food. For now though, we are sipping our first drinks and I have his attention, or most of it anyway. In the large pavilion to our right, well-dressed clusters of men and women are taking their places at tables around a dance floor—the wedding.

Mr. Njola was one of the architects responsible for introducing the prepaid LUKU meters into Dar es Salaam, and the festive atmosphere contrasts with his pessimistic reflections on the challenges of Tanesco. During my first interview in his home nearby, he observed:

So the issue here is the behavior, it requires someone to have financial discipline. Now, whether the financial discipline is brought by Tanesco or is inborn, lack of financial discipline will give you darkness and embarrassment in society, and next time you'll spend wisely for everything. In the past the behavior of Tanzanians, they didn't have that financial discipline, and it was like that because even the thief could go to the hospital and be treated for free. Why should he be worried about the finances? All the social services were free. With that in their head, people weren't cost-conscious, and that hangover is still there today, and if they are not cost-conscious, naturally they won't be financially disciplined. I think you've seen people in town, how they throw money around, how they throw booze everywhere. If you go to Europe, no one will buy you a beer [laughs]! It's not that they don't have money, they just have the discipline. So in Tanesco it's the same thing, and our customers are just the same, since we are still in the same society. Our efficiency as Tanesco depends on the society. Our customers, we can't even identify their names, how do we transact? One day a customer comes in and says his name is this, tomorrow he'll change his name. Then it comes to whom you are dealing with. You are dealing with nobody. Who are your customers? You don't even know who you are really transacting with . . .

Once fuel is turned into electrical current, it flows down to the city's bristling mass of individual households or businesses, where it passes through the meter and is measured in kilowatt hours. This is the first step in the billing and payment process that channels currency back upward toward the centralized Tanesco offices. But this is a critical "chokepoint" (Carse et. al 2019), one subject to all manner of parasitic interference. As Mr. Njola suggested, Tanesco customers are a kind of floating population. From the well-to-do suburban fringes of Mbezi Beach to the older, poorer neighborhoods, urban residents find that they are willing or able to consume more electricity than they pay for. This misalignment between the assumptions built into the meter and the tendencies of urban households can be traced back to the pressures of jobless growth and irregular flows of income that impinge on Global South urbanism as a whole. In Dar es Salaam some of these structural pressures were evident as early as the highly segregated colonial period, and despite CCM's attempts to resolve them in the socialist era of nation-building, by the 2000s they had only metastasized.

This chapter explores the negotiations that develop around this infrastructure of billing and payment from the perspective of those state agents charged with enforcing it. For three months, often accompanied by my friend, roommate, and research assistant Thierry, I conducted participant observation in two Tanesco regional offices in Dar es Salaam.[1] For some stretches we remained inside the office, assisting in the customer service and billing departments. Mostly, however, we were moving about the city in the back of a pickup truck with various field operations, including meter installation crews, a disconnection (DC) team, and a revenue protection unit (RPU)—the latter two tasked with disconnecting customers for debt, theft, or other meter irregularities. As seen with Mr. Njola, Tanesco management experiences these irregularities as deserving of social and moral censure. But out in the field, they were partially tolerated by Tanesco workers. In fact, as I will show, the real moral abdication was to act stupid or blind to compromise, to the forms of modal reasoning that allowed residents to access *some* measure of electricity and Tanesco to sustain some measure of cost recovery in difficult circumstances.

During my time with Tanesco I followed patrol teams across roughly two kinds of landscapes of domestic power use, ones that reflect the centrality of property ownership in shaping Dar es Salaam's morphology (Brennan 2012, 9). One cluster of electricity users—those in poorer, denser, and older core "Swahili" neighborhoods (*uswahilini*)—are distinguished by low levels of consumption, but high levels of debt that shade into illegal reconnection. These neighborhoods comprise the historically "African" areas of the colonial city and were subject to a modicum of electricity provisioning in the early developmentalist decades of independence. Starting in the 1980s, the original residents who settled in these areas began renting out rooms to waves of newcomers who were seeking income opportunities and looking to evade the increasingly strangled commute to and from the city's peri-urban edges (John et al. 2020). In recent years, these centrally located neighborhoods have seen some degree of gentrification, with landlords investing in infrastructural improvements to raise rents and attract better tenants, or with investors purchasing plots to build *maghorofa* (multistory buildings), particularly in the Kariakoo area (Andreasen and Agergaard 2016). On the whole, though, they are still characterized by relatively poor renters whose consumption and investment strategies are at odds with the type of modernist consumer-citizen Tanesco is premised upon. And they experience Tanesco patrols much in the same way their predecessors in such neighborhoods historically experienced state agents, as vectors of random and punitive intrusions; a kind of machete slicing through the thicket every so often, but not nearly often enough to clear the terrain.

Another stratum of domestic power users—richer, newer, spread out—is paradoxically both more and less agreeable to Tanesco's billing and payment operations. While there is no specific name for these neighborhoods, they are generally suburban, middle class or wealthier, and tend to self-consciously define themselves in opposition to *uswahilini* neighborhoods, which they characterize as dense and disordered (Mercer 2020, 530–33). With their titled plots, clear physical demarcations, and prepaid meters, they are in some respects more legible on the surface but more deeply opaque. While their prepaid meters preclude onerous debt buildup, these homeowners have more incentive to engage in premeditated forms of theft, meter tampering, and bribes. Theirs is a world ensconced—behind high gates, behind layers of guards and domestic servants, but most of all, behind money and influence. They embody the fortressed urban life that Victor Azarya and Naomi Chazan (1987) described as a strategy of "self-enclosure."

Tanesco patrols thus trace out a schismogenesis of slums and enclaves, the divergent geographies of socioeconomic disengagement from the state familiar across the neoliberal globe. And yet these polarized landscapes have hidden affinities, beginning with the fact that residents in each can find incentives not to pay for state power. For many renters, committing to paying for a constant flow of electricity is hardly the most prudent option in a situation of general volatility—there are much more reliable ways to spend one's money. For their part, landlords and homeowners may find that more sophisticated schemes for stealing electricity are much to their advantage. Small business operations or the accouterments of middle-class modernity—refrigerators, air conditioners, computers—require a lot of power that may be prohibitively expensive. In short, it is often the case that the power consumption of poor "Swahili" people is too low, while that of the middle and upper class is too high. This situation creates a complicated problem for Tanesco, which imagines its ideal customer as a modernist, atomized, and legible household with regular demand and flows of income.

While the colonial and socialist state had long been in the business of aggressively disciplining the population in the name of economic development, it was only with the neoliberal reforms of the early 2000s that consumer electricity came to be a plotline in this national drama. Up until then Tanesco was a subsidized parastatal of minor significance, operating on soft-budget constraints. Now it required a police wing to actually enforce its monopoly over a demographically and geographically expanding customer base. Tanesco was thus thrust onto the stage of state-society relations, and replayed many of the old dramas. On the one hand, with a new commitment to debt collection, Tanesco teams rolled through neighborhoods and physically removed wires, cut-out fuses, and meters: an aggressive scene that echoed earlier state crackdowns on the urban poor. For older Tanesco fieldworkers and managers alike, indebted customers attracted a wider state of antipathy toward parasites. During my fieldwork, this often took the form of diatribes against "Swahili" people, a rich if murky term of abuse for the urban poor, one deeply rooted in the colonial/modernist experience.

But Tanesco's late entry into the work of state discipline marked its increasing limitations. Tanesco is one of the most consistently unpopular government institutions, due to the unpredictable power supply and the way it indexes postsocialist forms of elite self-enrichment. Thus, while Tanesco patrol workers could strike a paternalist, colonial condescension against the irresponsible poor, they were often on murky terrain, both literally and morally. Residents can very well flip moral critiques about responsibility back onto an underperforming Tanesco. In the same way, patrol workers might try and strike a righteous socialist note with the rich, but often as not the rich "win the game" anyway, obstructing the teams with money, gates, or influence.

Tanesco itself had to endure the unofficial dispossessions that belied the official, rationalizing discipline of neoliberal "air-conditioner" reforms. The commercialization component of the World Bank's "standard model" (see chapter 1) mandated the retrenchment and hiring freezes of the Tanesco workforce itself. The thinking was that a digital prepaid system would automate billing and collection, and make redundant the messy face-to-face encounters between citizens, meter readers, and patrol teams. It has not quite worked out this way. Prepaid meters were implemented unevenly, leaving approximately half of Dar's population with the older, postpaid billing system. This system still requires meter readers to physically go to houses and monitor the meter, and it requires DC teams to manually inspect and disconnect faulty or tampered connections. Tanesco also still needs "boots on the ground" for its revenue protection, emergency, and service line crews. *Vibarua*—temporary workers on precarious short-term contracts—staff many of these departments. Even without hiring freezes and temporary contracts, Dar's population would likely outpace the number of required Tanesco workers.[2] Thus, even full Tanesco staff tend to be demoralized and resigned. They lack manpower, equipment, trucks—and, they feel, adequate compensation. They watch ambivalently as underpaid temporary workers, some of whom they've trained personally, leave "for the street" and leverage their skills and connections as *vishoka*—a phenomenon covered in-depth in the next chapter.

In turn, they enter into entrepreneurial *vishoka* activities out in the field as well, accepting customers' money to not disconnect them or forming quasi-protection rackets with them.

Field patrols are tasked with a number of difficult goals. They must execute their professional mandate to cut off indebted or stealing customers, but they cannot push their authority too far since they are often an underequipped, obstructed, and generally unpopular sight in Dar's neighborhoods. Finally, they must weigh these difficulties against their own compensations and satisfactions. Officially, Tanesco field teams "follow the disconnection sheet." In good high modernist fashion, they scold customers who are not properly "on the grid," who are connected through debt, informal sociotechnical arrangements, or outright theft. But outside of this bright formal spotlight, one can reconstruct a "hidden transcript" (Scott 1990) of surreptitious reconnection, quasi-protection rackets, bypasses, meter tampering, and the overall thriving of *vishoka* in different niches of Tanesco's institutional operations. In short, field teams cut those who can be cut, defer to those who cannot, and tacitly forge deals with those that fall somewhere in between.

These compromises highlight how street-level electrification is sustained under conditions of structural strain. Rather than fading into the taken-for-granted background of daily life, the consumption/payment circuit is continuously (if tacitly) suspended and diverted, delayed and restored. In the short to medium term, these arrangements allowed residents to consume more electricity than they might officially be able to pay for, and both salaried and casualized Tanesco workers to pay themselves. As I will show, these arrangements are modal—they concern the logic of *how* the consumer-utility relation can be modified, of the variations internal to its general coherence. This is an emergentist view of infrastructure as a living relation that can (and indeed must) wax or wane as it encounters parasitic interference. In Dar es Salaam, conflicts over "parasitism" date back to the late colonial period.

swahili urbanism

In 1957, four years before independence, Tanganyikan colonial administrator J. K. Leslie composed a sort of quasi-ethnographic survey of Dar es Salaam, in which he asked:

> Why does a man come to Dar es Salaam? To get money. For what? That depends—there are many who come to get capital, to invest in a wife, land, tools of a trade, a bicycle, goods; there are also those, a large class, who

come for a career, at rates of pay which enable them to reach and maintain a permanently higher standard of living. But the biggest class, drawn from the uneducated, unambitious, coastal tribes, comes to simply get spending money (quoted in Lewinson 2006a, 472).

Upon its formation as the center of German and later British colonial administration, Dar es Salaam was divided into three zones: a residential and administrative White area on the coast line and peninsula jutting into the Kigamboni harbor (*Uzunguni*); a high-density, exclusively African settlement to the southwest (*Uswahilini*); and a physical and social "buffer zone" inhabited by a predominantly South Asian commercial and residential population (*Uhindini*). In the interwar period, the level of physical infrastructure mapped onto this hierarchy: the White area featured wide, tree-lined boulevards, cooling ocean breezes and electrified buildings. The South Asian commercial zone benefited mainly from paved central roads and public streetlighting, often for security reasons. The African areas, though occasionally benefiting from settlement upgrade projects, remained unelectrified, as it was thought that no Africans could afford it. To some degree the evolving city spilled over these barriers. As early as the 1930s, Asian landlords began to purchase properties in the market area of the African zone (Kariakoo), while local Zaramo and Shomvi peoples sold their plots and moved ever outward to the edges of the city. Nevertheless, Dar es Salaam's structure as a "city of three colors" has fundamentally shaped its growth and ambience (Smiley 2009).

For colonial administrators, the challenges of governing *Uswahilini* coincided with the challenges of Swahili-type people. An *mswahili* connotes a lumpen, faintly outlandish, scuffling, and fast-talking personality. This reflects a shifting historical emphasis of the term, which was originally used by newcomers, slaves, and dependents (usually men) in coastal city states in the precolonial era to confer an aura of urbane respectability. "Swahili" was a quasi-ethnicity associated with conversion to Islam, literacy, participation in community rituals, and a distantly Arabic genealogy (Middleton 1994). However, the cosmopolitan, ocean vistas of the term took on a less salutary light in the race thinking of colonial discourse. "Swahili" came to stand for "detribalization" and miscegenation, for the non-industrious parasite awash in the commercial economy; the fluid coastal Other to the solid up-country migrant, whether the prosperous, missionary-educated Chagga and Haya to the mountainous northeast or the stalwart Sukuma and Nyamwezi in the central western plateau. It is this distinction that Leslie draws on when he speaks of "coastal tribes" interested only in spending money.

As Mr. Njola explained, Tanesco perceives one of its biggest problems to be the spending habits of its customers. As a modernist institution, it needs to convert current to currency in an even, regulated rhythm, and the fact that Tanesco does not have such a precise rhythm is often attributed to the moral failings of the populace. An editorial about Tanesco put it in much the same terms: "There is an inbuilt culture among many of us to shirk the responsibility of paying for services or goods provided to them. Whether it is education, health, transport or any other service, such persons would only pay if they are goaded. But, when it comes to contributing such superfluous things as weddings, send offs, or graduation ceremonies, many of us are ever so willing to come forward and outdo others in the amounts they give or at least, pledge."[3]

To colonial minds like J. K. Leslie and his intellectual inheritors, regularly paying for services or goods indexes *investment* in a "permanently higher standard of living," characterized by modern housing and amenities. It indexes entry into the formal sector in the broadest sense of the term, of modernity as a permanent, linear "structure of time" itself (Mbembe and Nuttall 2004, 347). The *mswahili* by contrast is a "spender-type" who pursues superfluous luxuries and leisure. His profligacy brings him no discernible returns and sticks him in the eternal present of tradition. For both Leslie and the writer of the editorial, this behavior is explained away by an "in-built culture" or "coastal mentality."

> True to the spender-type's aims in coming to town he puts [his salary] not into food or accommodation but into display or into leisure to enjoy the sights and sounds without the necessity to work every day. More than half the dockers are Zaramo or Rufiji, both with the coastal mentality well developed: this mentality is not based on ambition, as is that of many of the up country tribes, but on the ideal of the old Arab coconut plantation owners, in whose systems of priorities leisure and a Hamlet-like freedom of choice ranked high, while effort is a minor consideration. (quoted in Lewinson 2006a, 473)

Such attitudes have formed the ideological underpinnings of state contempt up until the present day. However, one of the central achievements of economic anthropology has been to problematize rigid distinctions between investment and spending, consumption and production (Guyer 1997). Such distinctions are ill suited to understanding the economic logic of rural peasantries and large, poor cities alike, which fundamentally depend on a more holistic sense of monetary *circulation*. As one Tanesco worker beautifully phrased it, life in Dar es Salaam is like the tide, sometimes it takes you to shore, only to drag you back out to sea. And since most people lack steady incomes to get them to shore each month, this means they must press upon those who do. In his classic work on informal economies, Keith Hart (1973, 65) pointed out that the "manipulation of the credit system" is a classic way to swim this tide—to negotiate the "chronic imbalance between income from wage employment and expenditure." For example, he describes how Frafra migrants in Accra would form "chains of debt" whereby pressure to repay creditors would be transferred to one's own debtors, in infinite regress (Hart 2000). Like Kula valuables or other gift exchanges, money must move. It acquires its value through circulation, by constantly deferring the moment when accounts are settled.

Circulation is the wellspring of the substantivist tradition, which underemphasizes money as a thing-like commodity and foregrounds it as a sign, a token or measure of social relations (Hart 1986; Scott 1977). And while this is true, the fact that money can be stolen or diverted or otherwise transferred also means that it is a material medium with particular affordances. The Kiswahili word to express this clever appreciation for the possibilities of mediation is *ujanja*, or ingenuity. Irregular income—compounded by the pleasures and aspirations of urban life—also tempts people to take calculated risks and shortcuts. Dar es Salaam is rife with everyday gossip of people who don't repay loans, who take advantage of a family member's kindness, trick customers, and overcharge strangers, or just outright cheat a little bit on a transaction. For lives lived on tides of money that sometimes bring you to shore and sometimes take you out to sea, these kinds of short-term gambits can be as reasonable as the collective insurance mechanisms of long-term social reciprocity. Both respond to the fact that the modernist regimentation of economic life "can be dangerously irrational for individual actors," partly because within "conditions that people expect to shift at any moment" it would be a mistake to allocate money to something that does not immediately require it (Guyer 2004, 164).

Whether as solidarity or *ujanja*, circulation helps to explain why meter debt tends to build up in *uswahilini* households. What if something more important comes up?—a relative gets sick, a car breaks down, or a loan repayment doesn't come through? Moreover, regular electricity is often relatively low on the list of household expenditures in *uswahilini* neighborhoods. At the high end there are the necessities of social reproduction such as rent, food and water, transport, and cooking fuel (charcoal, wood, liquid natural gas, kerosene), as well as loans and some kinds of pressing repayment. Next there is the cost of children's education. This would include not only the big expenditures in August and December, when school fees are due, but also regular daily money for lunches, transportation, etc. Obviously, the budgetary importance of electricity changes when it's part of a small-to-medium enterprise such as a bar or grain-threshing operation, but it would seem that otherwise residents are trying to cover a number of other expenses each month before they get to paying the power bill.

This is not to say that households are profligate power consumers. In general, poorer households are correlated with lower electricity use in Tanzania (Shibano and Mogi 2020), and tend to reduce already low levels of electricity consumption to prioritize more pressing expenditures when necessary. Households in *usawhilini* have long foregone electric stoves that were briefly adopted in the 1990s, before the tariff spikes of the NetGroup era, and are quick to unplug freezers, turn off televisions, or hold off on ironing to minimize expenses (Ghanadan 2008, 164). Still, *umeme* (electricity) is hard to eliminate completely, and any ambition to purify its flow into zones of productive and unproductive use is bound to falter. When I conducted a survey of fifty households in three *uswahilini* areas (Vigunguti, Tabata, and Buguruni), some male respondents pointedly noted that it was women who, by virtue of spending more time in the house, tended to use more fans, radio, or television during the day. Others referenced norms of "provider love" (Hunter 2010, Stark 2017) that oblige men to furnish their partners with just such amenities, though of course men are hardly immune to the comforts of a cooling breeze or diverting broadcast. A complex mixture of need, desire, and entitlement—often gender inflected—animates what we might call socially necessary electricity consumption.

Against this background, perhaps the most important reason that electricity debt builds up in *uswahilini* is the simple fact that the billing and payment system affords it. Tanesco's post-paid system, like any credit system, buys the consumer time. In principle that time extends only to the end of the month, and Tanesco's customer service charter (and engineering manual distributed to employees) reserves the utility's right to discontinue service for any amount of unpaid debt thereafter. In practice, there seems to be a fairly long-running norm of extending that grace period, with both official and unofficial variations across branches and regions. In 1992, for example, the World Bank commissioned a Power Loss Reduction Study for Tanzania that, much like my own fieldwork twenty years later, included interviews and participant observation among various sectors of the utility. There the author wrote: "[T]he head office in Dar es Salaam stated that when each new bill is issued, disconnection orders are issued for accounts that are one month's payment in arrears. The Morogoro office reported that disconnections are undertaken on accounts that are three months in arrears. The observed facts did not support either assertion. Many cases were found where active accounts were much more than three months in arrears" (Hay 1998, 6).

While accompanying patrols, Thierry and I found that DC teams never really concerned themselves with any debt below Tsh.100,000 (see Degani 2021), nor did any household I surveyed report being disconnected for a debt below Tsh. 100,000 (the lowest was Tsh. 120,000). When not on patrol, I worked the front window in the Tanesco billing office to process payment and reconnection procedures (see figure 3.1) and saw that many customers took advantage of this lag—or at least passively accepted it. Every day we would print out consumers' monthly bills and hand them out to the crowd, most of which were in the Tsh. 50,000–100,000 range. The wide variation in household composition notwithstanding, my household survey found that the average household electricity bill is probably Tsh. 20,000–50,000, suggesting that people came in to get their bill around every two or three months. We might thus infer that many residents are "paying close to the edge," and taking advantage of this tacit buffer. In this limited sense, Tanesco's conventional billing system is retrofitted to the economic logic of informal incomes, a logic that constantly defers full payment.

This is not to say that residents in *uswahilini* prefer the conventional metering system. To the contrary, they overwhelmingly prefer the prepaid "LUKU"

FIGURE 3.1
The author hands out
electricity bills at a
Tanesco branch office.

(*Lipa Umeme Kadri Unavyotumia*—pay as you use) system,[4] which provides a far greater amount of control and customization of electricity consumption, allowing people to buy in small increments more characteristic of their income flows, and automatically shutting off the power when they are out. The popularity of LUKU piggybacks upon the great technological success story of neoliberal Africa—the rise of prepaid telecommunications networks. As my friend and fellow anthropologist Mohammed Yunus Rafiq once pointed out, with their keypads to input purchase codes and digital displays, they even sort of look like (older) mobile phones. This is a telling contrast to prepaid meters in their original setting of South Africa, where they are widely seen as a disciplinary, anti-poor technology borne out of the apartheid regime's intention to combat a politicized "culture of non-payment" (von Schnitzler 2013; Van Huedsen 2009). Crossing the Tanzanian border in the late 1990s, prepaid meters were seen as a liberating technology. Their social "script" (Akrich 1992) did not imply irresponsible consumers who wouldn't pay their bills, but proper citizens who should enjoy autonomy from all the hassles and inefficiencies of the state (Jacome and Ray 2018), as befitting the new, "air-conditioned" iteration of developmental modernism. As one resident succinctly put it to me, with LUKU *giza kwako*—the darkness is your own.

However liberating the technology, its allocation was still subject to Tanesco's socialist-style practices of hoarding and shortage. The relative scarcity and expense of LUKU meters means that they are reserved for new connections and the occasional wholesale upgrade of wealthier neighborhoods. Only after these prioritized customers may households apply for a LUKU replacement. Most of the older neighborhoods that were electrified before the 1990s still retain the conventional payment system (Kiangi 2015), and thus the *sine qua non* for massive amounts of debt buildup.

Being stuck with a conventional billing system is particularly ill-suited to the broader dispensation of electricity consumption that has accompanied prepaid meters, one marked by a mixture of deteriorating service and increased expense. The 1960s–1990s was the era of highly subsidized electricity provision, with industrial-to-residential cross-subsidized tariffs and a 1,000 kWh free "life-line." In response to the precipitous devaluation of Tanzanian currency during structural adjustment, Tanesco passed the cost of power onto the consumer, unraveling cross-subsidies, cutting the lifeline down to 50 kwh, and implementing marginal cost pricing (Ghanadan 2008, 79–84). Additionally provoked by Tanesco's emergency style of management, electricity costs are fundamentally inflationary. Prices were hiked 40 percent in 2007, again in

2012, and again in 2013, as discussed in chapter 1, exacerbating the possibility of debt buildup among even the most economized households.

Conventional meters also encourage debt through the messy procedures of reading and billing. The aging conventional metering technology is in fact rather imprecise. Residents cannot easily keep track of how much they are using, or may just forget without an explicit reminder in the form of the digits on a LUKU display. Frayed cables or other technical problems may fool their general feel for it, which contributes to transmission losses. Tanesco is chronically short of trucks and bodies, and perpetually behind on a disconnection list that spans pages and pages of indebted customers. The possibility that they will find and disconnect any particular meter in a timely manner is far from certain. Alongside political scandals and widespread load shedding, these factors combine to breed popular mistrust that further disinclines residents to "disciplined" bill payment.[5]

On the consumer side, one of the biggest reasons residents in *uswahilini* have for distrusting the modernist framing of economic life lies in electricity's ambiguous relation to the household. Swahili-style houses feature individual rooms that line a long narrow hallway or border an inner courtyard. Many are *nyumba za urithi*, or inherited houses that landlords rent out to tenants, while they themselves may or may not live on site. However, the compound as a whole tends to have a single meter, which causes complications. By paying collectively, many households are disqualified for even the reduced lifeline tariff (50 kwh, provided use is < 236 kWh/month), increasing the financial strain. Parsing out who is responsible for individual use is also very difficult. In 2004, a survey of neighborhoods in Sinza and Manzese illustrated some of these dynamics. Out of two hundred houses only one was satisfied with a meter-sharing arrangement, and they had essentially formed their own mini-distribution company, complete with designated tariffs on specific appliances, payment dates, and borrowing procedures (Ghanadan 2008, 164). On this topic, one Tanesco clerk, a born-again Christian named Mr. Chogogwe, remarked to me editorially that such disputes might be especially venomous in Muslim compounds. Multiple wives prone to jealousy and deceit exacerbated the tensions of multifamily housing. Of course, this stereotype need not be restricted to Muslims. It is hardly unheard of for Christians to practice de facto polygamy.[6] More generally, "the image of the gossiping, two-faced combative Swahili housewife [runs] through urban culture" (Lewinson 2006a, 478). If nothing else, such gendered caricatures intimate the frictions of close-quartered living.

The landlord-tenant relationship is an equally fraught hazard of urban life. Both sides must proceed cautiously with each other, exemplified in the common practice of paying upfront for six months to one year of rent. For landlords, this preempts rent default, though it is unquestionably a financial burden for tenants. But it also prevents the landlord from arbitrarily raising the rent "if he sees you living well," as one resident explained. These ambiguities also extend to electricity. Because there is no standard arrangement as to who is responsible for paying electricity bills, exploitation is always a possibility. On the one hand, an unscrupulous absentee landlord might collect money for the bill and then pocket it (as happened to a friend of mine living in Temeke). On the other hand, the landlord is vulnerable to tenants who rack up debts and abscond. A landlord once came into the Tanesco billing office and actually requested that his house be disconnected. His tenants had racked up a debt in the hundreds of thousands of shillings, and he was afraid he would be "left with the ball" (*kuachewa na mpira*). A few days later, one of the tenants came in and begged to be reconnected, as it was destroying her grain-threshing business. Her fellow tenants, she added bitterly, were of no help. Because Tanesco legally sides with the landlord, she was gently told that nothing could be done.

These practices mean that electricity debt often forms a residue of the city's shifting human geography. The modernist presumptions of the conventional billing system involve a kind of *tabula rasa*: a household corresponds to a set social unit, has clearly delineated responsibilities over payment, and the means to meet that payment. However, many houses in *uswahilini* host cycles of renters—or, in some gentrifying areas, buyers—who unwittingly inherit unpaid debts on the meter. An interlocutor, who gave himself the magnificent nickname of "Dunky No Position" and whom I will refer to as Dunky, explains:

> In order to avoid problems with Tanesco, it's said that when you want to move into a room as a renter, you must first ask: "does your meter have a debt on it?" You can't just conclude "oh, it's a nice house, why don't I just rent a room?" Lo and behold: the meter has a debt. At the end of the day Tanesco will read the meter, and at the end of the day they'll come and apprehend you the renters. You need to ask the landlord: are you indebted to Tanesco? It's your right to ask. But many don't ask and many run into problems. At the end of the day people are crying.
>
> This happened a lot in Kariakoo. In the past there were normal houses [there]. After the owner sees his debt has gotten too large, he sells the house to a white person, an Arab, a Chinese [person] . . . whatever. The buyer

doesn't think about this. The day they cut the power he goes and sees he has a debt of Tsh. 20,000,000. He'll try to bulldoze the house in order to build a multistory building, and he'll be told "the number on this lot registered a debt and we can't connect the power until the debt is paid." So the debt's now left with you, the stranger/guest (*mgeni*). And the Swahili has already gotten his money. He goes to build a house out in the bush, out in Chanika. And that's that.

When residents add up all the disincentives to play by the rules of regular electricity bill payment—when their meters are opaque; when understaffed meter readers don't often make it to tucked-away houses in unplanned areas (and when the tedium of multiple handwritten data entry practices easily produces billing mistakes); when the possibility of disconnection seems relatively remote but also often arbitrary—it becomes easier to see how debt builds up among consumers to such extraordinary levels. And then equally important are all the incentives to play by the alternative rules of the popular economy. When renters are only provisionally tied to their housemates and landlords, and when economic expenditures tend to privilege "investment in social relations" (Berry 1989) in the form of loans or gifts or celebrations, we can see such investments both downgrade electricity bills in terms of budgetary allocation and eventually convert them to their own logic of deferred payment. Unlike Tanesco, kith and kin can keep close tabs on residents, and their relationships are far more dimensional. The scope of Tanesco's sanctions for default, while it does have a certain coerciveness, is usually limited to disconnection. By contrast, low-impact shaming, pestering, ostracism, gossiping, and partial excision from reciprocity relationships are all part of the repertoire of interpersonal credit.

a short history of state raids

I have shown how the social, economic, and technical landscapes of *uswahilini* create a situation in which debt can build up very high. I now want to offer some ethnographic reflections on what happens when Tanesco DC teams enter such landscapes.

It is the second day of patrols and the truck has dropped off two members of the DC team in a Sinza neighborhood, a man named Chilambo and a woman named Sabina.[7] The terrain is hilly and even boasts pleasant tree cover, a testament to its desirable and well-established location. Though near the built-up core of the city, the place still has *mpepo*, or air circulation. We have entered into a kind of clearing where four or five large houses are perched on a gently descending slope. In front of one house, two young girls, approximately nine and fourteen, are washing clothes in a blue plastic tub and then hanging them up to dry on a nearby clothes line. They are, we soon see, the only ones around.

For the first twenty days of the month, DC teams are assigned to track down meters with outstanding debt. During the last ten days, the teams conduct general inspections. This is to give disconnected households a chance to begin the reconnection process before the start of the next billing cycle. But often teams will use this period to double back on houses they have recently disconnected, giving their patrols a sort of staccato rhythm. Today the redundancy has paid off. Sabina explains: "They were cut two days ago. They had a debt of Tsh. 7,000,000. At least they could've reconnected through the meter—at the very least. But they've dodged the meter entirely. Better to get charged for having your power cut and reconnecting it, rather than getting charged and just stealing. Behold Tanzanian life. This is the problem . . . [Tanesco] is going down because of the stupidity . . . the ignorance of not paying one's bills."

Disconnections follow a general script. When a household is to be cut off, a DC technician will explain the situation and yank out the small black cut-out fuse with a satisfying pop. The action then slows to a bureaucratic plod, as two separate copies of the disconnection form are filled out by hand. During this time the customer will plead, argue, stall, or—as we will see later—proffer payment to "resolve the issue" right then and there. Inspectors are meant to counter these ploys as best they can and move on. The contrite customer is meant to bring one copy to the nearest regional office and pay all debts and reconnection fees.

Of course, there are plenty of exceptions. The team immediately saw that the household in Sinza had decided to take matters into its own hands. Stretching out from the utility pole, the wire ran around the meter and connected directly to the household circuit, a rough and ready kind of job usually done by a local kishoka. In cases of premeditated theft, the minimum response is to cut the wire itself. However, perhaps due to the magnitude of the debt, or the audacity of stealing power just two short days after the patrol came through, Sabina decided to take legal measures. The plan was to pursue a kibali (license or warrant), and return with the police later that night to make an arrest. As Sabina explained these measures, the two young girls began inching back toward the house. Seeing this, Sabina called to the older one, playfully. She took out a small digital camera from her bag, lined up the two girls and snapped pictures of their faces, wide-eyed with fear. Laughingly, she taunted them: "Ala, c'mere!

It was a sweet day for you renters today, you felt cool when using the easy power [i.e., the air conditioner]. Have you ever slept at the police station? The landlord's name is what? You don't know? Our police are young . . . you're the one caught in the house, and so you have to go. Tonight they'll be sucking on your breasts [*leo watakunyonya matiti*]!"

I was taken aback by Sabina's cruel threat and her bizarre enjoyment in delivering it. Clearly she was angry at the brazenness of the theft, and saw it as a particularly egregious example of customer "stupidity" about not paying bills. But consciously or not, Sabina was also enacting a scene with deep historical resonances—one that ultimately implicated my own presence on patrols that day, and one that dramatized the intrinsic "infrastructural violence" (Rodgers and O'Neill 2012) of disconnection, whereby access to public services is differentiated by class, gender, and ethnicity (e.g., Anand 2012).

There has long been a culture of state contempt for populations not plugged into the circuits of the "productive" formal economy. In chapter 2, I discussed how this contempt took shape around modernist ambitions and nationalist discipline, but its roots stretch back into the late colonial period. In the postwar period (1945–1961), the colonial state feared rampant urban population growth and its attendant street economy as sources of criminality and even political dissent. They embarked on a policy of labor stabilization, attempting to provide full employment for educated workers while excising the "unproductive" poor from the city. This principally took the form of tax raids and "spiv campaigns," designed to repatriate ostensibly unemployed idlers to outlying village areas (Burton 2007). But these campaigns could hardly keep up with Dar's inexorable population growth. Despite the state's welfarist and interventionist ambitions, swaths of Dar es Salaam remained within, but not of, the striated space of state order. As Brennan (2008, 99) succinctly puts it: "Forms of urban colonial governance, from land titles to urban migration controls to tax collection . . . did not infuse daily life through representative structures, but struck only on occasion and with sharp force from unaccountable outsiders." Thus for the majority of Dar es Salaam's colonial-era residents, a central scene of state power was extradition, a rupture into an otherwise segregated "African" space of *uswahilini*. It is no wonder that this period was awash in rumors of *mumiani* (vampires) hidden behind bizarre state uniforms, dark-windowed cars, and walled-off state compounds, waiting to strike (White 2000; Brennan 2008).[8]

Municipal authorities in Dar es Salaam have continued to ritually "sweep" urban spaces, in spite—or rather because of—the limited results. In the socialist era, unemployed youth were "repatriated" to collective villages through campaigns like the 1972 Operation *Kupe* (Tick); the 1976 Operation *Kila Mtu Afanye Kazi* (All Must Work); and the 1983 Operation *Nguvi Kazi* (Hard Work) discussed in chapter 2. Throughout the 1990s and 2000s, authorities repeatedly bulldozed informal markets and harassed mobile *wamachinga* (street vendors) that wended their way through the wealthy, historically "white" parts of the city. The most recent addition to this public menace is the swarm of young motorcycle taxi drivers that thread the city's ever-jammed road network. In the 2010s, an urban planner charged with developing the latest master plan professed his admiration for Singapore's authoritarian capitalism, replete with extensive slum clearance, massive infrastructure projects, and relocation to "satellite towns" (Boyle 2012).[9] The state thus continues to nurture high modernist ambitions. But, as in the colonial and socialist eras, its power is top heavy,[10] reduced to dramatizing these ambitions in occasional—though certainly damaging—raids and inspections.

I listened to Sabina's threats with this history of state paternalism and aggression—and its limits—very much on my mind. Walking away, a team member named Chilambo laughed that it was highly unlikely that they would be coming back with the police anytime soon. For that kind of assistance, the Tanesco security office would be required to pay "gas money" for the police to come at night, when the responsible parties were likely home. Even then, as the head security office later explained, the resident might be able to bribe police or the courts to run the case aground.

Second, night raids were an inherently sketchy proposition. Like Americans' proverbial dread of IRS audits, DC teams are an unwelcome, portentous sight in Dar neighborhoods. Combined with Tanesco's notoriously bad service, blackouts, and expensive prices, patrols could often feel like they were venturing into enemy territory. Indeed, residents often greeted DC teams with grumbling resentment. One angry customer declared "you are all *vishoka*, you just want money," while another referred to us as *popo*, or bats for the strange way workers "hung" from utility poles. As I was unfamiliar with the word, *popo* was explained to me as that strange animal that's both a bird and a beast, since it has wings but also has milk. In other words, like *mumiani*, *popo* are classic matter out of place, and perhaps fitting for skilled state employees who, since the colonial period, have moved through local communities and engaged in a species of aggressively mysterious labor.

Moreover, residents could occasionally strike back with violence of their own. The manager of the Revenue Protection Unit, for instance, bitterly recounted how he had been punched in the face and had once even had a gun pulled on him.[11] A few weeks later, I learned that before he worked in the comfort of the air-conditioned billing office, Mr. Chogogwe used to go out on DC patrols. During one particular night raid, as the police were dragging away the offending debtor in the dark, the man's son hurled a rock at Mr. Chogogwe's head, knocking him unconscious. The DC inspector who told me the story said he understood the outburst. How can one just sit in the darkness and allow his father to be taken away?

These incidents cast a fuller, more strategic light on Sabina's overt aggression toward the two young girls in Sinza. Since the responsible adult was out in the city, there was "no better way to compel him to pay his bills than to threaten his daughter," as Chilambo put it.[12] Sabina's threat may have indexed the history of state raids, but it was just that—a threat. Moreover, the situation was exceptional because it was a relatively secluded encounter with children, and it involved a large debt and premeditated theft. Usually, the DC teams dealt with the lower-stakes procedure of disconnecting households for smaller debts or technical irregularities. More often than not, patrol teams found themselves on the backfoot, tangled in the complex webs of distributed ownership and responsibility. As I wrote in my field notes: "To determine if a customer had done anything wrong was to cut through a maze of excuses, of inherited debts, of other actors not on the scene, of absentee landlords and tenants late on the bills, of misplaced documents and technical malfunction, of spatial ambiguities, of ignorance feigned and real."

And usually multiple tenants, neighbors, and passers-by were around to protest or talk back. At a cramped pharmacy attached to a residential compound, for example, a meter's broken security seal drew Sabina's suspicions and an argumentative crowd. Amidst the back and forth, one man came up and said the house belonged to his brother, a lorry driver who was not often around. How, he demanded, could Sabina possibly imagine that the remaining *akina mama* (womenfolk) would have any idea about technological things like a meter? (The irony of asking a woman inspector this question did not seem to register.) In another neighborhood, two women, visibly swallowing their anger, begged an inspector named Mary not to disconnect them, explaining that they had recently moved into the house and unwittingly inherited a Tsh. 500,000 meter debt.[13] Deciding the simplest course of action was to pay it down themselves, they signed an agreement at the Tanesco office to pay it down in installments in exchange for a new LUKU meter. They were a mere Tsh. 33,000 behind on payments and in a week or two they might have cleared it. But Mary, for the simple fact of being there that day, was obliged to disable the LUKU. To be so marginally over the line was a bitter pill, though she took no evident pleasure in the act either.

MARY: These renters were left with the debt, but here we appear as the bad guys. If you have compassion you can't do your work. . . . If they could just pay even like 50,000 they would get their power back. But sometimes customers are convinced you are harassing/oppressing them.

MZEE MCHUZI: You can't argue with them whether they owe or not. You can explain that the responsible party is the landlord, not you. There's an agreement: you rent, the landlord takes your money and pockets it. Tanesco charges for electricity. When we come to cut the power we're cutting off the landlord, not you the renter. It's the landlord who owes. And so [when] a person says "it's not me who owes"—that's none of our concern.

MARY: If it were otherwise we would be going around to each house asking: is this debt the landlord's or the renters?

MD: I see that lots of customers get angry and complain.

MZEE MCHUZI: [laughing] They enjoy electricity but don't have the means to pay.

MD: Do you find it hard to do this kind of work because you have to accept all sorts of harsh language?

MZEE MCHUZI: [laughing] It's hard, yes. But what can you do? It's the work we have.

MARY: And you can't fight the customer. You just do your work and move on, because every time you try and explain he won't understand it. You can't fight. You just follow your disconnections sheet.

Inspectors are, as one once described to me, the "police wing" of the utility. In reality, they are at best partially allied with actual security forces, and their ability to move through neighborhoods was constantly beset by all manner of friction and contestation. My own presence, I think, seemed to indirectly illustrate this predicament. One hot afternoon, when Mzee Mchuzi commented appreciatively on the quiet demeanor of a woman whose meter had just been disconnected, Sabina laughed with satisfaction: "that's right, because the bosses are around." While not literally true—Thierry and I were officially "trainees," relegated to the role of gophers and ladder-haulers—there was a structural truth in her remark. As a young *mzungu* (white) male foreigner with a well-dressed Tanzanian assistant, both of us out of place amidst the middle-aged workers in green Tanesco coats and grey work pants, I was obviously connected to a larger economy of "bosses"—of managers, consultants, World Bank "loss reduction specialists," etc.—that stretched back to the J. K. Leslies of the colonial era.[14] Indeed, one sickening thought I've had over the years is that my presence might have *emboldened* Sabina to threaten those girls, inspiring her to invoke the power of a modernist state with all of its white colonial overtones and histories. For despite the intrinsic "infrastructural violence" of their power to disconnect, it was also clear that inspectors were usually outnumbered, undersupplied, and more or less unable to "fight the customer"—a relatively democratizing condition that spawned all manner of compromises and complicities.

too many eggs

For inspectors, the inability to "fight the customer" is exacerbated by Tanesco's own institutional thinning in the post-NetGroup era. Consider an episode from my first week with the DC team. Coincidentally—or not—a supervisor from Human Resources had been assigned to accompany the inspectors that week. Joy had just graduated from the University of Dar es Salaam and had plans to pursue a master's degree abroad. Until she could save up enough money, she was working at Tanesco, a difficult to acquire position suggesting the help of personal connections. Ostensibly, she was going on rounds to forge closer connections between the office management and field workers, though it may just as well have been to keep an eye on the field team and/or the anthropologist. Out at the "site," as it is called, Joy proved disinterested in the goings-on of patrols, often tiring quickly and electing to rest in the shade as the team members went about their inspections in the vicinity.

On our third afternoon that week, at a moment when we were all moving together, we came face-to-face with another technician walking toward us. He was a young man in the classic green Tanesco coat. No one seemed to recognize him. After exchanging greetings, Sabina appraised him and called out half-jokingly: "that uniform belongs to me." Without quite stopping, the technician laughed and replied, "No, it's not yours, but it's good for getting up onto the rooftops."

As he walked away, Joy asked after him. Mzee Mchuzi explained it was probably a *kibarua* (temporary worker) who had somehow gotten hold of a Tanesco coat to "get onto the rooftops"; that is, to impersonate fully salaried employees enfranchised to meddle with meters and wires. In other words, he was a *kishoka*, the very sort of unsavory element known to reconnect cut-off meters or, alternatively, scam customers. Suddenly affronted, Joy demanded he be found and made to return the coat, but Sabina cut her off: "don't make a case out of it for yourself, don't go searching for trouble." Mzee Mchuzi added flatly, "Don't be surprised, they are everywhere, sometimes even with official Tanesco ID cards. If people can steal entire [import] containers, how hard is it to get a uniform? Sometimes they know more than we do."

It is rare to see a direct encounter between *vishoka*, management, and workers, and more interesting still for the way it reveals the usually invisible, institutional forces that shape patrols. In his discussion of labor in the Zambian Copperbelt, Ferguson proposes that mid-twentieth-century mining, for all of its exploitation, "entailed a very significant broader social project. It was . . . socially thick" (2006, 197). This included housing, amenities, and educational programs. In interviews, Tanesco inspectors offered similar sentiments. They described the different ways that the company had invested in them: they received uniforms and soap to wash them once a month, some were given company housing (with free power), and were taught many different facets of their trade. Mzee Mchuzi started working for Tanesco in the 1970s, and spoke with pride about the old policy that had circulated him through construction, emergency, installation, and transmission. The principle, he explained, was to create a generalist workforce that "knew every part of the work," and that could be presumably allocated more effectively. One cannot also fail to notice how this accords with the Nyererean policy of transferring civil servants around the country to uproot any "tribalism" (*ukabila*) and create a truly nationalist cadre.

Today, thanks to consumer growth and institutional reforms, this social thickness has diminished. DC inspectors are a kind of heirloom or "legacy" workforce, remnants of what the Left historical tradition sometimes calls a labor aristocracy (Hobsbawm 1984; Parpart 1984), squeezed by the twin pincers of technocratic management from above and casualization from below. As Chilambo observed:

> You can't compare a secretary or manager that gets Tsh. 2,000,000/month whereas as an employee you get Tsh. 600,000 but you did the lion's share of the work . . . we grapple with power and do the work without rubber, without gloves, and that's why there are many technicians who lose faith and insert themselves into cooperation with *vishoka*. Because [they] know a lot about the work but don't get anything in return. [They will] say "how am I going to sleep beat-tired after doing wiring from morning to dusk, when my coworkers are only leaving the office?"
>
> That's how they combat the difficulty of life, by getting this kind of [*vishoka*] work. Like today we cut [a household with a] debt of Tsh. 1,000,000. But there's going to be a person there [in the neighborhood] who knows

electricity well and will return their power, so he can get something to help him out in his life. [The problem is] that Tanesco is in the habit of laying a lot of eggs but they don't have the capacity to rear them. We create so many *vibarua* [temporary workers]! Personally I've already taught more than one hundred *vibarua* but now they are all contractors and fine technicians who work in the street [*mtaani*] and know electricity. If Tanesco would employ these people who know the work these sorts of problems wouldn't happen. I've taught one young man named Kasi, with a fine intelligence, finer than mine. He's worked in Arusha, Mwanza . . . and if he says he's tired you know he's really tired! A person like that, you think he'll just sit around for Tsh. 500/day?

Chilambo thus sympathizes with electricians that are trained by Tanesco—indeed that he himself helped train. Why would they want to work in dangerous and poor conditions when their education potentially entitles them to something better?

regulating desire

The flourishing of *vishoka* and the ambivalence of the salaried Tanesco workers that initially train them speak to broader ethical predicaments in postsocialist Dar es Salaam. One day after rounds, Chilambo reflected: "There's something called self-confidence (*kujiamini*) that bedevils us. And even today it entered. I went to cut the power and was supposed to wear gloves, but I just went like this [holds up bare hands]. But that electricity . . . [shakes his head]. And why? Because of self-confidence. It's an issue that needs to be addressed educationally, that you shouldn't have excessive self-confidence."

Self-confidence is part of a broader semantic field anchored around the term *tamaa*, a kind of libidinal energy that might simply mean desire but also ambition. It is associated with hot forces such as money, alcohol, or lust, which, because they are generatively potent, must be properly channeled (cf. Weiss 1996; Myhre 2017). Writing of Kaguru communities in Eastern Tanzania, Thomas Beidelmen (1997) describes ritual and moral education as a "cool knife" that cuts the heat of desire. More broadly, the ethnographic record is also full of accounts of youth who strive to "tame" the money they make (De Boeck 1998), converting it into long-term socially productive assets such as cattle, mosques, or remittance-financed housing construction (Ferguson 1985; Fioratta 2015; Melly 2017). It is also full of accounts of the opposite: "grasping individuals [who] divert the resources of the long-term cycle for their own short-term transactions" (Parry and Bloch 1989, 27). Francis Nyamnjoh has observed that "far from denying or downplaying the existence of animosity, hostility, aversion, or conflict," popular and scholarly African thought often thematizes "social life as a contested terrain of tensions . . . needing a careful balance of intimacy and distance" (2017, 264; see also Vokes 2013, 102–5).[15] Vital substances, which can regenerate or ensorcell, express this double edge of desire.

Moreover, in a spirit of "conviviality" (Nyamnjoh 2017, 267) we can find elective affinities with traditions of "Western" social thought that have little truck with a unified subject or stable social field but rather emphasize aporia and ambivalence. As individuals, Emile Durkheim suggests, we are irrevocably *duplex*, forever trying to align our appetites with "our attachment to something other than ourselves" ([1914] 2005, 36; McGovern 1999). We are here back at what Mazzarella (2017), reviving this predicament in the paradox of mana (expressing both sacred order and insurgent charisma), calls the dynamic interdependence of "social form" and "vital" energy, or what the Wittgensteinean tradition thinks of as the interplay of form and life (Das 2015). Without the energy to substantiate it, the formal premises of social order remain hollow; without form to channel it, vitality remains mere potential or worse—think uncontrolled fire or escaped electrical current. Indeed, in Deleuzean terms, there is always something deterritorializing about too much desire. It is evident in the "infuriated" pack, swarm, or crowd, in which the social body is dissolved into waves of energy (Galloway 2013, 60–62), or (closer to the case at hand) of the "Swahili" layabout constantly running around, pulled this way and that by appetite and circumstance (Degani 2018). And yet by the same token, for residents of Dar es Salaam, to "cut desire" (*kukata tamaa*)—a synonym for giving up—is also desubjectifying. It is in its own way to stop thinking, connoting as it does a kind of stupefied, bestial submission to one's fate.

The postsocialist period might be understood as one in which new energetic desires were pumped into the country without the infrastructures to properly realize them as social form. Thanks to public budget cuts and lifted price controls in the 1980s, many urban dwellers felt chronic "cash hunger" (*njaa*), as both staples and luxury goods frequently remained out of reach (Lewinson 2003, 17). These reforms tended to exacerbate the steady decline in real wages that had been unfolding since the mid-1970s (Tripp 1997, 40).[16] As the cost of living grew faster than what salaries could meaningfully cover, entry swelled into an already substantial informal economy that, depending on a variety of factors, might entail high-risk, high-reward scores (crime, smuggling, scams, and cons) or drip of daily sustenance (*wamachinga* vendors, petty entrepreneurship, urban farming, contract labor). Each mode—illicit, illegal, unofficial—had its tradeoffs and belied a range of intermediate options and links to more formal kinds of employment. Each provided a different kind of answer as to how residents could avoid being reduced to their socioeconomic position or, conversely, to their desires for something beyond it. Each was a way of configuring the ongoing circulation between their desires and the world around them.

On the one hand, inspectors spoke in disapproving middle-class tones of consumers and *vishoka* whose reach exceeded their grasp. Chilambo reflects: "I've set myself up in the city, I've married, I educate my children and do everything. To do that you have to respect your work. But they [the youth] say 'I want to build a *ghorofa* [multistory house], and I want to buy a car' when even I, who started back in 1983, have never had a car. I don't even have a bicycle. I respect the work because I support my family." At the same time, they also understood the structural conditions that thwarted those desires. When asked if he would ever consider leaving Tanesco, Chilambo answered:

> Because of the age I've reached [probably not], but perhaps it could happen. I'm only human and humanity isn't careful all of the time. We live by luck, like bed bugs. You know bed bugs? Even if they don't bite you there's still a smell. We could do nothing wrong and a person could say "ah he's a thief" and [the management] says "oh he's a thief eh?" And then you can just get fired, because not everyone that gets shut down has done something wrong. Our security is that of bed bugs. So anything can happen, and if that happens—fine, you have to accept it. And then you have to ask yourself "where can I go?" There are lots of private companies now that have tried to entice me but I haven't acquiesced to them because I don't see the reason, because this is the place that has floated (*kuelea*) me since I was small until today, and I've married and raised children, so I've gotten used to it.

Chilambo is a "company man." He was lucky enough to secure a position within what was a socially thick labor regime in the 1980s. He states proudly that he has never been accused of any corruption, takes pride in his work and in the young technicians he has trained. But he also recognizes that this stability is increasingly subject to a number of forces he cannot control. He might lose his job at the hands of a vindictive customer. He might suffer injury from his own lapse in *utaratibu*—the proper, methodical way of doing things. More prosaically, the very technicians he trains may turn against Tanesco itself.

These wild forces demand the right mixture of endurance (*kuvumilia*) and self-confidence. While Chilambo is oriented toward the former mode, walking the narrow line of legitimate employment, many "youth'" have neither the luxury nor the disposition to do so. In the next chapter I discuss the career arcs of street electricians and Tanesco temporary workers that enter in *vishoka* activities. Like Chilambo, they aim to convert the energy of youth into marriage, house, and family—the lasting fruits of social reproduction. But here we see them from the perspective of Tanesco workers, as they pass each other in the field. *Vishoka* are a sort of "fifth column," but one whose presence Chilambo understands, even sympathizes with. They may be the vectors of illegal reconnection and meter tampering, but like the DC team, they are born of Tanesco itself.

cutting without cutting connection

Absent the proper resources or incentives to categorically enforce billing and payment, patrols could seem hollow, almost formalist exercises. No matter how many households the inspectors cut, they were all too aware that the disconnection sheet would continue to grow, and that many households would simply surreptitiously reconnect, often through the very *vishoka* they might have trained. Indeed, sometimes the work of a single disconnection required an almost comically large expenditure of energy relative to the daily quota of households (Degani 2018). As I elaborate in this section, doing nothing more— but importantly, nothing less—than following the disconnection sheet is a more complex task than it might first seem. It involves the sometimes subtle but ultimately crucial difference between cutting and "cutting connection" (Degani 2021)—between treating nonpaying customers as difficulties to be tolerated and treating them as enemies to be expelled.

Consider first the particular method of disconnection. Despite its paeans to customer discipline, Tanesco would rather retain a nonpaying customer than lose them altogether. Even with a household with relatively high levels of debt, the DC inspector will simply yank out the cut-out fuse rather than unhook the wire or confiscate the meter. Full disconnection of post-paid meters effectively cuts off Tanesco's own blood supply, reducing the chances that an offending customer will pay without Tanesco pursuing legal action. Like threatening (but not actually jailing) a man's daughter, simply removing cut-out fuses is a prudent compromise. It punishes the customer, but not so harshly as to alienate him or her from the relationship entirely. It is a kind of modest gift or last chance on the part of Tanesco. It is a grace period, a suspension, a limbo: it cuts without cutting connection.

Such grace is certainly vulnerable to consumer abuse. Once residents are disconnected, at least three logical possibilities open up to them. Chastened, they may simply pay their debt at the office. But as we've seen, they may also completely bypass their meter and steal power, as did the household in Sinza. The third option is the most interesting for the way it mediates both extremes. Residents can pay a *kishoka* to obtain and reinsert a cutout into the meter, thereby resuming the status quo ante in which the meter continues to record debt. This arrangement ideally buys the consumer time. It may continue for months or even years before it is discovered, given the scant coverage of DC patrols and the limited training and loyalties of subcontracted meter readers. The DC teams favor staccato rhythms of return on their patrols for just this reason—recent illegal reconnections are more likely to be caught by them than by Tanesco's stunted bureaucratic antennae.

If the reinserted cutout does manage to buy residents some time, they can use it to collect enough money to pay down their debt without having to lose power and pay a reconnection fee. Alternatively, they may elect to run up the bill and let the unpaid debt ripen into a kind of retroactive theft. Still, as Sabina observed of the house in Sinza, such "theft" would at least be recorded on the meter, thereby preserving the pretense that it might be paid back. In a sense, this tactic mirrors and redoubles the phatic logic of the original disconnection, whose aim is to preserve the social relation by suspending contact and thus forestalling the debt spiral. The response then *suspends the suspension*, bringing everything back to first position. This circular, almost polite *fort-da* of removal and reinsertion expresses a kind of "agonistic intimacy" between household and utility, a "picture of relatedness predisposed to neither oppositional negation nor to communitarian affirmation" (Singh 2014, 171). That is, it allows both actors some short-term latitude while remaining oriented towards a long-term horizon of reciprocity. As disturbing as Sabina's threats were, it is significant that she perceived them as symmetrical to the residents' own thoughtless negation of their future together.

An alternative to reinserting the cutout, which allows consumers to continue racking up debt as if the DC team had never passed through, is to pay the DC team to look the other way when they actually do come through. But this act itself has a tacit logic to it. Dunky, who has worked as a Tanesco *kibarua*, offered the following fee structure for bribes (*rushwa*) to avoid disconnection:

DUNKY: It depends how much he has on him inside the house. You can't say "go to the bank and I'll wait for you." But yes, sometimes he'll say "I don't have any inside, come back later and I'll give you Tsh. 500,000." You go ahead and later you'll come back.

MD: Let's say for example I have a debt of Tsh. 5,000,000. A DC worker comes, we talk, and I say "fine I'll give Tsh. 150,000" [to not disconnect me].

DUNKY: Impossible, that's too small an amount.

MD: Because I have such a large debt . . .

DUNKY: Yes.

MD: If I offer Tsh. 300,000 . . .

DUNKY: Impossible. If you have Tsh. 5,000,000 [in debt], you have to arrive at least Tsh. 800,000 or Tsh. 900,000.

MD: Ok if I have Tsh. 10,000,000 in debt, I would have to pay at least how much.

DUNKY: For Tsh. 10,000,000 . . . we have to cut your power [laughs]. That's a lot of money . . . 10,000,000 is a calloused debt [*deni sugu*], a debt that has fully ripened [*deni liliokomaa*]. So rather than tell him to give you Tsh. 200,000 or Tsh. 300,000, it's better that he just pays it over [at the office].

MD: Let's say I'm a normal person, I live in Buguruni and only have a debt of Tsh. 300,000 . . .

DUNKY: Go pay at the office. Those small debts of Tsh. 200,000 or 300,000, that's money that after a month or so you could go and pay. Better I cut so that this month a paycheck can cover it. There's other money that is alright [to take], Tsh. 600,000 or Tsh. 700,000. You give me Tsh. 300,000 [of that], no problem.

Dunky's outline suggests that there is a "sweet spot" for which a payment to prevent disconnection is acceptable, somewhere approaching Tsh. 1 million and at an upper bound of Tsh. 5 million. Like reinserting the cutout, such payments allow consumers to keep accruing recorded debt. They also have the additional benefit of allowing Tanesco workers to supplement their income. Small-scale debts of Tsh. 100,000 or 200,000 are too small to accept payment on and too large not to cut. Large-scale debts are also too large not to cut, but conversely, they are too large to service through bribes. In effect, this structure prevents poor and wealthy households from consuming stolen power to excess while tolerating a midrange of negotiation.

As moderate solutions, mid-range bribes and reinserted cutouts remain within the general logic of the distribution network, but residents' decisions to employ them are in no sense dictated by that logic. This is not a functionalist argument, in other words. As I saw on patrols, residents routinely engage in excessive forms of theft and workers are more than capable of extortion and abuse. The point is that by local estimations they are acting selfishly and foolishly. Confiscating the entire meter that day in Sinza, Sabina could well understand the household's impulse to continue using electricity. She was insulted that they could not, evidently, restrain that desire by reciprocally seeing themselves from Tanesco's perspective. As she said herself, if they were going to steal, "they could've at least reconnected through the meter—at the very least." But alas, they had flouted even this courtesy and were stealing power directly.

Sabina's use of the term "at least" reflects a kind of modal reasoning. She does not specify how much worse it is to illegally reconnect through the meter compared to paying one's arrears and then reconnecting. But she establishes that these two acts are in a different category from the direct bypass, which does not in any meaningful sense modify the sociotechnical link between consumer and utility but rather pushes it to the breaking point. It completely diverts power, choking off all upward flow of payment and, by extension, the elementary criteria of the relation.

As Michel Serres (1982) might put it, the failure to restrain one's desires is often a failure to see one's self as part of an ecology. In *The Parasite*, he imagines the problem of hares in a farmer's garden, advising the latter, "Don't chase the hare out; you would need the entire armed forces to do so. . . . [Say] good morning, hare; stay if you like. As long as you have one hare in the garden, only one hare, it's better to make your peace with it" (1982, 88). In other words, Serres distinguishes parasites as either a kind of nuisance or noise one must coexist with, or as an *enemy* that must be expelled through armed force. The latter is a fantasy of purification that is ultimately self-negating: the only way to completely eliminate the hare is to level the garden itself.

The high-modernist fantasy of eliminating social and economic parasites is one the Tanzanian state has long indulged (Brennan 2006a). It often literally treats such figures as enemies, deploying militarized operations (*operesheni*) to sweep urban space of quasi-legal street vendors, motorcycle taxis, and other informal operators, not to mention electricity thieves—as Sabina's threats illustrate. Despite these militant campaigns, the "parasitic" informal economy has metastasized in Tanzania, as it has in much of the world. Its deterritorializing tendencies are evident across a wide range of phenomena in neoliberal Africa, from roving war machines (Hoffman 2011), to frenzied visa lotteries (Piot 2010), to xenophobic campaigns of "turning the clock back to zero" (McGovern 2012a). Coursing through all these practices is a desire and sometimes desperation to act in real time, a sense that survival or accumulation demands that actors "*bypass* the intricate questions of maintenance, ownership and so on" (De Boeck 2011, 271; emphasis mine). But the bypass is in some ways a fantasy in its own right, a mirror image of capitalist aspirations to pure liquidity unencumbered by social relations. As Guyer (2007) has written, the "near future" of the neoliberal imagination evaporates into a perpetual present of boom and bust, with its evangelical overtones of atomized winners and losers, the elect and the indebted. While this surely approximates the extreme reaches of African urban life, we might also attend to the way ordinary residents resist the evacuation of a near future by weaving together surfeit and suffering, shortcuts and restraint, disconnection and reconnection. When one is forced to play the wily hare or the vigilant farmer, it is useful to know when not to give chase. The phatic removal and reinsertion of the meter cutout is this thought materialized in infrastructure.

peri-urban landscapes and postsocialist households

Because they are charged with debt collection, DC teams stick to older neighborhoods that were electrified with conventional meters from the 1960s through the 1990s. By contrast, the citywide Revenue Protection Unit, which I was transferred to after about six weeks, looks for deliberate theft where it is most likely to find it: in the white areas, the gentrified commercial center of Kariakoo, and new peri-urban developments. Their patrol routes thus reflect the city's ongoing socio-spatial evolution. In the postsocialist era, the city's middle and upper classes staked out new patterns of housing and settlement (Mercer 2020), ones that repudiated patterns of life and money in *uswahilini* while simultaneously bearing the stamp of their influence.

In her article on domesticity in Dar es Salaam (2006a), Anne Lewinson charts the formation of a class of African home-owning professionals. In the socialist era, the National Housing Council was able to provide some housing to state bureaucrats and workers. These were rented flats in the international socialist style, with an emphasis on austere, atomized functionality. However, many professionals were not so lucky and were forced to rent rooms in the unplanned, mixed-income *uswahilini* neighborhoods. Professionals may not have liked their poorer, ostensibly traditional Swahili neighbors, whom they often privately characterized as gossips and busybodies, yet socialist ideals of egalitarianism mixed with indigenous domestic practices imposed a common sociality. Starting in the 1980s, many urban professionals living in *uswahilini* adopted informal, officially prohibited income-generating side projects in the home, such as renting spare rooms, farming, or raising animals to sell. Along with norms of generalized reciprocity—attending weddings or funerals, loaning money, watching children, using personal connections to help with bureaucratic obstacles—these activities pushed social classes together in everyday life, sometimes uneasily.

By the late 1980s, the economic shocks of structural adjustment had prompted many urban residents to purchase agricultural plots on the city outskirts to ensure adequate food supply (Briggs and Mwamfupe 1999). At the same time, rural migrants increasingly settled along these unserviced peri-urban areas, creating linear settlements clustered around arterial roads leading out from the city center. For such migrants, access to cheap plots, informal land-tenure mechanisms, and urban agriculture outweighed the difficulties of poor infrastructure (Kombe 2005). By the 1990s, deregulation of public transportation and the increasing availability of private cars had facilitated daily commutes into the city for work or trade (Rizzo 2002). The middle and upper classes began to populate large swathes of land between the major arteries. For white-collar professionals and state bureaucrats, housing was seen as an appealing market for any windfalls accumulated—sometimes dubiously—in the post-structural adjustment economy.

The result was a number of homogeneous, middle- and upper-class residential neighborhoods in the peri-urban areas on the outskirts of town. However, while geographically separate, these neighborhoods in some ways ironically reflect the Swahili classes that urban professionals were at pains to distinguish themselves from in the 1980s. Whatever their resentments about living in *uswahilini*, diminishing incomes had instructed them in the same informal logics of investment and expenditure. Lewinson (2006a, 490) observes:

> One could read the explosion in home construction as a sign of new prosperity—or as a sign that fired workers were investing their severance packages in something solid, namely a home of their own. In the second scenario, office workers were calculating that even if they could not find another job, they would have a rent free roof over their heads. Perhaps the choice to build a home or create several micro-enterprises rather than a major manufacturing business revealed a basic lack of confidence in the newest model of modernity, liberalization, espoused by the government. . . . Once

again, urban Tanzanians were creating an indigenized form of modernity that anticipated scarcity and contingency, rather than upward progress.

Thus, while signaling social distinction, home construction was also a relatively safe investment in an uncertain economic climate, and a difficult one for the government to (re)nationalize. A fancy new home may look like an unproductive vanity project—almost harkening back to the Arab Plantation ideal that J. K. Leslie derided in the 1950s. But as with other supposedly nonproductive assets (e.g., celebrations and displays associated with social reciprocity), it was a sound investment given the unpredictability of informal income.

The architecture of professionals' homes embodied these changing circumstances. In key ways, it contrasted to the classic Swahili-style house. The latter is a machine for economic production as much as for living: dense settlement patterns and open architecture facilitate social interaction and livelihood activities. It anchors vegetable gardens and storefronts, while porch space is often rented out to tailors or vendors. Local celebrations such as weddings, funerals, and *ngoma* (dance celebrations) take place among them, as do high volumes of regular socializing. By contrast, new housing was a status symbol that represented the lifted restrictions on individual consumption in the postsocialist era. Expanded living rooms and private sleeping quarters diminished the bathing and cooking areas and brought them indoors, thus providing only enough room for a single family. Finally, because it is such a dense concentration of "wealth in things," and because such neighborhoods are often tucked away and not easily accessible without a car or motorbike, they are enclosed by the ubiquitous wrought iron gates (produced with great skill by local welders) and thick concrete walls, often topped with glass shards or sometimes even electrified fences.

This ensemble of built form, economic logic, and geography influences patterns of electrification. The key distinction is that unlike renters and absentee landlords in *uswahilini*, homeowners are financially and infrastructurally "locked in" to sustained electricity use. In older Swahili-style houses, for example, cables and wires often form a web across the floors and walls, linked to single-phase electrical connection. By contrast, circuitry in newer houses is embedded within them and powered by a three-phase connection, a testament to the expectation of continuous lighting, refrigeration, television, internet, and mobile phones. Moreover, because many of these houses are built in unserviced areas, it is incumbent upon residents themselves to pay for the exterior poles, wires, and meter from Tanesco, with their own identities tied to the process—a complex set of transactional pathways that I examine in the next chapter. Finally, newer houses built by the 2000s almost all have prepaid meters, which ensures that consumers cannot rack up large amounts of unpaid debt.

The combination of high-volume power consumption with a billing system that cannot shade into de facto theft incentivizes premeditated theft. Wealthy residents can expect to devote a portion of their monthly income to refrigerated food, conditioned air, indoor lighting, television, and even computers. It is only reasonable that they will seek ways to mitigate the expense. Residents may hire *vishoka* to tamper with their meters so that it records more slowly, or create switches that toggle between metered and unmetered electricity, or set aside money to pay off inspectors should they get caught. Whereas a *kishoka* in *uswahilini* might reconnect a cut-out fuse for Tsh. 10,000 to 15,000, my interlocutors suggested that any of the former options will likely begin at Tsh. 100,000. Accordingly, theft comes to resemble a kind of long-term investment that requires a good deal of start-up capital.

Finally, each kind of illegal connection will shape the "cat-and-mouse games" with Tanesco patrols. The challenge for patrols in *uswahilini* is to move through social and spatial density. In principle, DC teams can move from house to house rather quickly. Meters sit on outside walls, or in houses that are only weakly differentiated from the immediate outside environment. On the down side, inspectors are often ensnared in arguments and the illegible, distributed nature of responsibility for the meter and payment. They confront a weakly differentiated network of landlords, families, and co-renters that do not always stay in one place for very long. Unless the residents in question happen to be wealthy or organized enough to offer a bribe, DC teams are more likely to just "cut through" this confusion and disconnect.

In wealthier neighborhoods, the challenge is obstruction. Meters are often locked behind a compound gate or thick door and, likewise, the social relations that embed them are more hierarchical and defined. Again and again on RPU patrols, domestic servants, guards, and relatives would assert that they didn't have the authority to let us in, as the house did not legally belong to them. Although the homeowner is ultimately responsible for debt or theft, such dependents were often very nervous about interacting with us. They rightly understood that *they* would be held responsible for admitting the RPU team in the first place. Hence we were treated to many solemn declarations that we could not possibly be admitted. This was a rather inane bit of theater, but one with a convincing set piece and mood lighting: an iron gate physically enclosing the space, and the RPU team's reluctance to press the issue in the face of Tanesco's unpopularity. In one such house, the RPU team discovered that a meter was rigged such that two of its three phases were not recording power. After they wrote out the disconnection form in duplicate, the truck arrived out front. I followed them out of the compound to help with the ladder. One inspector glanced back, jumped a little, and told me to remain where I was. "Why?" I demanded. I was instantly suspicious that some bribery proffer was afoot. "Because," she answered, "if we all walk past the gate, they'll probably lock us out."

building a certain kind of environment

If and when RPU inspectors do finally get access to the meter, and find that it has been tampered with, they may solicit or be offered payment not to disconnect. I have to tread somewhat carefully here and emphasize that I never saw any DC or RPU team member directly accept any money. Nevertheless, the issue demands consideration because while my presence on field patrols had an obvious dampening effect, I have certainly seen customers offer bribes, and there is a general perception that field teams are known to accept them. As Scott describes of his fieldwork on peasant resistance, "without ever pursuing the matter directly, a pattern of facts nevertheless emerged from casual listening" (1985, 265). For example, an apartment complex in the wealthy, downtown market area of Kariakoo only had one working meter for all the apartment units, presenting a serious threat of overload. The meter, moreover, was registered in Temeke, Dar es Salaam's southern district (Kariakoo is part of the Ilala district). After disconnecting the meter, I couldn't help but notice that an inspector gave the resident responsible for the complex his phone number—"just in case the landlord wants to speak to me." He then left to go make a copy of the disconnection paperwork. Curiously, we never actually made it back to the complex. Later, he simply dismissed the whole thing, saying, "you give these guys the paperwork and they never take it in anyway."

In the abstract, inspectors spoke of such sly deals in a way that strikes me as both evocative and precise. They called it *kujenga mazingira fulani*, literally "building a certain kind of environment" in which one can take money from customers. To build an environment depends on cultivating certain spatial and/or discursive possibilities. Inspectors must find "off-stage" slivers of ambiguous space, time, or language, away from prying coworkers, supervisors, or anthropologists. Evading the first was especially important, I gathered. In a moment of candor, one RPU inspector said laughingly that if you hear your coworker was accused of taking money from someone, you immediately feel jealous. You wonder why you weren't told. Surely, then, any RPU member involved in such transactions must consider whether having a coworker involved—for coordination, alibis, assistance, or the like—is worth sharing the money.

While inspectors are known to find interstitial moments of negotiation, away from prying eyes, Tanesco has posted bulletins that warn of a more baroque possibility, a sort of concealment squared:

Warning to Customers Concerning the Wave of
"Vishoka" Conmen in Various Areas of the Country.

The Tanzania Electric Supply Company (Tanesco) is announcing to all customers throughout the country that there is a resurgent wave of people who pose as Tanesco workers. These conmen have fake IDs and stolen uniforms.

These Conmen or Vishoka often come as if to inspect the customer's meter and fool the customer that his meter has been tampered with. After this fake inspection, they remove the "Cut Out" device that allows power to enter into the house. They then tell the customer that the price of the amount of power stolen is large and that they can reduce that debt by having him disburse an amount of money or they will report him to the authorities.

Tanesco requests all of its customers throughout the country not to accept this scam and [reminds them] that neither the Company nor its workers are permitted to accept any customer money outside of the Tanesco office. The customer should make sure that even if he does owe anything he goes to pay the debt or his payment at the Tanesco office and that he's given a receipt that has been printed out by the computer.

This will help to catch these Vishoka Conmen.[17]

The redoubled nature of the scam here should be noted. The trick hinges on the idea that customers would *expect* actual Tanesco agents to solicit bribes—but would at least provide some service in exchange (such as debt clearance). Thus, not only are the victims complicit in bribery, the bribery itself is fake. The problem of *vishoka* illustrates how dramaturgy and spectacle are not just reserved for premodern "theater-states" but flow through the heart of sociopolitical authority itself (Geertz 1980). That even the most bureaucratic state power must ultimately be performatively materialized (i.e., executed) means that an actor can inhabit the role with the right props, costumes, dialogue—and the right performative panache.

Of course, panache cannot be underestimated. Consider the following email containing an account of an encounter with *vishoka*.[18]

Dear Colleagues,

Yesterday around midday, 9 men invaded my compound in Mbezi Beach while I was at work. They claimed to have come from Tanesco HQ and were there to inspect the LUKU meter. The house was locked so they could not access the meter. They harassed my guard and threatened to take him to the police station if he did not call me. The guard called me and I spoke to the men on their phone.

The purported inspector told me that they had used a GPS machine to measure the activity of my LUKU machine, and found it defective. He told me that he would assist me to correct the problem instead of referring the matter to Tanesco Office, but he needed some money to do this. I told him my guard does not keep money and asked him to come to UNICEF. He and four other men came to UNICEF but refused to enter the compound. I and one of the KK guards went out to talk to them. They asked for Tzs. 3 million. I tried to get them to move closer to the gate so that people would see their faces but failed. They were very guarded and the vehicle they used had very dark tints on windows. They left after a few minutes.

I refused to give them money and they left, threatening to go and disconnect power at my house. I called Tanesco for advice and was forwarded to the security section. The security personnel advised me not to take any action before they got here. When they arrived, I narrated the whole story to them. They were quick to point out that the inspectors were fake. Tanesco does not use GPS to measure anything. The 9 men were con-men popularly called "vishoka" who have been terrorizing Mbezi, Ubungo and Kigamboni for money. He informed me that 6 such men were caught the day before and were being held at the Central Police Station. It was his guess that the men who came to my house and that of another lady in Ubungo were looking for money to bail out their colleagues, which would explain their aggressive behavior. I gave them the phone number of the person who called . . . and the number plate of the vehicle.

Please warn guards and helpers at your residence. The group barging into the compound will comprise not less than 6 people. They must be prepared not to open the gate and must call for help immediately. If the men make it into the compound, it will be virtually impossible to control the situation. They have strength in numbers. They are loud, aggressive and very well-dressed (and well-fed!), with authentic-looking Tanesco ID cards, forms, cameras etc. There was even one man bearing what looked like BBC ID, representing the media.

On a positive note, my guard is unharmed. I was a bit traumatized by the whole experience.

The circumstances of this scam are somewhat unique since they involve a presumably white expatriate. But they demonstrate how, even with all the right props, a "one-note" performance of state authority can still fall apart under close scrutiny. Indeed, had these men been on an official RPU patrol, it is hard to imagine that they would have been so aggressive. All too aware of their limits in the field, inspectors (both DC and RPU) told me again and again that you "cannot fight the customer." You cannot force the customer to let you in, and you cannot force the customer to accept being cut off. Presumably, you cannot force the customer into a protection racket relationship either. If the negotiation is too one-sided, if RPU teams push too hard, it might backfire—something the *vishoka* in question learned the hard way. Whether negotiating a real bribe or coaxing a fake one, the ideal situation is one where customers consent to the alternative arrangement. Without some minimal compromise and complicity, forceful expropriation is an unsustainable basis for the inspector-customer relationship.

However, it would be a mistake to draw a hard distinction between "smash-and-grab" conmen *vishoka* who trick customers and legitimate Tanesco inspectors who negotiate with them. As I explore in the next chapter, *vishoka* often do have some connection to Tanesco itself. For instance, it is unlikely that the props of the escapade in Mbezi Beach—the truck, the uniforms (though not the GPS)—were easily acquired. Someone in Tanesco likely provided the cars and uniforms, and/or perhaps some were full Tanesco employees themselves.

Though Tanesco management portrays *vishoka* as impersonators, categorically outside Tanesco's main institutional channels, they are well aware of their blurring. It is not unheard of for RPU patrols to trick customers out of their money—the very thing that Tanesco attributes to *vishoka* "conmen" who cunningly pose as employees. At the first morning RPU meeting I attended, Mr. Bahari, the head, office-bound RPU manager, singled out two inspectors in particular. Apparently, a customer had come in complaining that she had given two RPU workers cash to reconnect them. They pocketed the money and then *kaingia mitini*, disappeared into the trees, never calling her back. She was shown a lineup of pictures and identified them as the culprits.

Bahari threatened that if they pulled anything like that again he would fire them straight away. Throughout my time with the RPU, he often began morning meetings with similar warnings and threats to the team as a whole.

Ultimately, whether performed with greater or lesser finesse, bribery points to the ways an entrepreneurial logic has come to animate Dar es Salaam's professional householders and Tanesco workers alike. Whereas white-collar professionals might have once been able to self-identify as the rightful beneficiaries in the national imaginary of state provision, the circumstances have changed. Lewinson observes that "since the 1980s . . . they have survived from informal economic activities rather than wages. Contrary to their expectation, they were not 'living [well] like an office worker' in a bureaucracy-centered city. . . . This reality clashed with their deepest hopes and expectations of city life" (1998, 218).

The patrols of the RPU inadvertently trace out the spatial and ideological disidentifications that have shaped this class. Instead of living in tidy government flats, they now invest in their own privatized forms of long-term security—the house on the peri-urban outskirts. Writing in the mid-1990s, Lewinson suggests that such residents invested their dashed hopes with a tragic depth and anger. Two decades later, they may find themselves in a more ambiguous situation in which their life projects have managed to cycle along in spite of themselves. As a class they have survived, in some ways even thrived—but the substantive mechanisms of social reproduction have shifted. Steady salaries have given way to the pervasive informalization of their economies and lifeworlds, and public scripts have been hollowed of their referential content even as there is meaning in the performance. In short, they might well ruefully concede that in some ways they have grown more like their Swahili neighbors, who have long understood the *ujanja* required for urban survival.

conclusion

In certain ways, the morphology of postsocialist Dar es Salaam is an intensification of trends well underway since the late colonial era. Whereas that period saw formal spatial segregation between Swahili neighborhoods and the planned settlements of the white-cum-African elite areas, the contemporary period sees a de facto split, with those same Swahili neighborhoods now equipped with an aging postpaid system and the city's elite having fled to suburban satellites, tethered by the digital umbilical cord of a prepaid metering system. The properly modern resident is still contrasted with the lazy and wayward Swahili resident, and there is still a dual-tiered labor hierarchy of full-time salaried employees and the informal knockabout—here embodied in Tanesco inspectors and the street electricians and dayworkers below them. The socialist period ameliorated these trends—most *uswahilini* neighborhoods near the core have some modicum of infrastructural provisioning—but on the whole their trajectories obtained.

Much of the scholarship on neoliberal Africa would suggest that these continuities are deceptive, reflecting not much more than the ossified shell of a certain kind of developmentalist ideal. If mid-twentieth-century modernist developmentalism (and particularly its socialist variants) was characterized by an orientation to the future that advanced from tradition to modernity, then, as Ferguson (2006, 177–78) points out, this axis of ascendance in time is now tipped onto its side to form a "horizontal" logic of membership. Thus, rather than the collective uplift of rural peasantries and urban workers, there are those who are in, and those who are out. As with the xenophobe and the alien of reactionary politics, or the saved and the damned of charismatic Christianity, there is the peri-urban enclave and the urban slum, and only miraculously shall the one become the other. In many ways, Tanesco's cost-recovery strategies first forged in the NetGroup era reproduced this split. The financial indiscipline of "Swahili" residents consigned them to permanent disconnection, while the luxury afforded to new householders was that their prepaid meters welcomed them to consume power any time (they had the money to do so). The logic of slicing up populations in terms of service provision, itself related to large trends in socioeconomic differentiation, has helped create "the different worlds of Bongo," as writer Mlagiri Kopoka puts it (chapter 2)—the neoliberal worlds of being definitively off or on the grid. Can they be bridged?

Indeed, this chapter argues that underneath these apparently calcified dichotomies there are a number of interesting convergences. First, the urban middle and professional classes have grown to resemble their poorer coresidents in a common "indiscipline" or "culture of nonpayment." The poor Swahili class may not practice saving or responsible payment on their power bills (and for generally intelligible reasons), but the schemes of power theft that proliferate through the wealthier compounds mean that as customers, the latter are not necessarily any more reliable. Second, the boundary between full-time salaried workers and their casualized counterparts is sanded down. Some full-time salaried Tanesco workers themselves become part of this infrastructural manipulation, striking out on their own to form their own entrepreneurial deals and hustles, even as they condemn the indiscipline of customers or *vishoka* electricians who contrive similar sorts of arrangements.

Thus, in the face of structural pressures wrought by liberalization, residents, *vishoka*, and Tanesco employees all experiment with ways to modify and thus sustain the otherwise diminished and irregular flows that nourish them socially, whether of current in the case of urban households, or cash in the case of Tanesco's labor force. A poor household might surreptitiously reconnect their indebted meter such that they are essentially billing themselves for illegally consumed power; a rich household might pay to alter the wiring or meter such that only a fraction of total use is billed, or bribe Tanesco workers to look the other way. Though often illegal or at least unofficial, at their best these modifications can have a functional quality. Temporary theft can act as a kind of debt refinancing, while bribes might be thought of as a tax on the rich or a labor subsidy to state workers whose wages have been artificially compressed by structural adjustment. They are modal in the sense that they qualify but do not quite break the larger collective commitments embodied in consumption and payment of electricity. They *strain*—in the linked senses of stressing and sieving (Kockelman 2017, 139–42)—the continuous movement of current and currency.

To be sure, one can get too clever with modifications. Taken to its logical conclusion, incrementally manipulating the circuit threatens to drift into the mere pretense of commitment, with patrols and residents constantly cutting and reconnecting in a continuous game of cat and mouse. And sometimes, conversely, the game falls apart. Actors take too much or play their position wrong. They find themselves foolishly or greedily reduced to their desires. Past a certain point, residents are not simply deferring payment: they are negating it. Tanesco workers or *vishoka* aren't paying themselves: they are unduly exploiting their clients. In such moments, social dramas ripple out in the wake of Tanesco's patrols, their recriminations and impotencies reflecting the latent contradictions of Tanzania's developmentalist ambitions in an era of liberalization.

Finally, this chapter may also be seen as the third iteration of a theme explored in this book, namely how certain emergent forms within Tanzania's national ecology are both preserved and undercut in the wake of liberalization. Thus, chapter 1 used the provisioning of emergency power to trace how the seeming continuity of CCM as the enduring center of Tanzanian political life belied a number of substantive admixtures within it, such as the embrace of "Asian" capital and decentralizing, oligarchic networks partially restabilized through rejected privatization bids and anticorruption sweeps. Chapter 2 explored how the experience of an urban or national public was constantly frittered away through power cuts, and partially restabilized through announcements and signification more generally. And in this chapter we see that the assumptions of what it means to be a proper urban household requires a whole series of tacit, fiddling negotiations that state workers must tolerate and delimit. From the perspective of what postsocialism means as a historical condition, perhaps we get a view of Tanzanian society admitting complex changes through the sociocultural backdoor: the substantive composition dilutes—but does not quite dissolve—the nominal form. The next and capstone chapter is a final variation on this theme. It turns to the career arcs of *vishoka* and explores how *vishoka* act as channels that mediate the process of connection to the grid. Their presence exploits the maintenance and extension of the grid, yet in some ways works to sustain it.

And how necessary caution is, the art of dosages, since overdose is a danger. You don't do it with a sledgehammer, you use a very fine file.

—GILLES DELEUZE AND FÉLIX GUATTARI, *A THOUSAND PLATEAUS*

4   becoming infrastructure

VISHOKA AND SELF-REALIZATION

This last chapter turns to the modal compromises that residents contrive around network maintenance and extension, and in particular of their local service lines. It is appropriate to end with the network's service lines, as in many ways they are the most crucial element of the entire system, the "last mile" that links the centralized public side of the network to its bristling mass of private, decentralized endpoints. After years of disinvestment, these narrow umbilical cords fray, and the process of extending new ones is bogged down by bureaucratic complications and delays. This adds another more basic layer to the problems of high tariffs, low supply, and faltering revenue collection that beset the production and consumption ends of the system. Without the clinching service line, no communication (however strained) between consumer and utility is possible.

Enter *vishoka*, those fixers who make a living on—and off of—the service line connection. In the previous chapter we briefly met a *kishoka* while working a neighborhood with a Tanesco Disconnection Team, and heard tell of some others—the aggressive con artists who attempted to extort a rich expatriate.

As these examples suggest, *vishoka* is a fuzzy category. It can refer to street electricians or licensed contractors, Tanesco day workers or ex-employees. For a price, they can provide replacement parts or labor for downed service lines, or expedite applications for new ones. As demand increases for this kind of under the table work, it has been supplied by a reserve army of underemployed, skilled, and semiskilled labor created in the neoliberal restructuring of Tanesco described in chapters 1 and 3. In many ways one can see *vishoka* as a covert—and covertly tolerated—reconstitution of Tanesco's labor force, in which formally underemployed actors find ways to "pay themselves" by providing supplemental services. They are "in" but not entirely "of" the network, and in this ambiguous sense illustrate the ways both infrastructure and personhood have become more dynamically experimental in the postsocialist era. I call the intertwining of these processes—the progressive incorporation of *vishoka* into the network—becoming infrastructure.

personhood as infrastructure

While infrastructure may conjure up visions of material substrates, ethnographers have attended to the ways such systems are inevitably peopled (Star 1999; Zárate 2018; Carse 2019). Workers are a particularly vital part of postcolonial waste management, transportation, and other support systems—their activity routines just as important as capital inputs such as roads or trucks, and indeed often compensating for the latter's deterioration (Fredericks 2018). A focus on labor highlights the centrality of practice, subjectivity, and skill in allowing complex sociotechnical systems to function, as well as the important role of larger political-economic conditions in feeding or starving those systems. As suggested in chapter 3 and further explored below, the restructuring of Tanesco's workforce has had important ramifications for how current and currency flow in and out of households.

A related but diverging branch of scholarship in urban studies and geography begins with AdbouMaliq Simone's (2004b) idea of "people as infrastructure" more generally. Originally used to describe the hustle and bustle of downtown Johannesburg, the concept was quickly recognized as a rather brilliant shorthand for the way urban life in the Global South seems to work despite its ambient dispossessions and poverty (Wilson and Jonas 2021). Part of the appeal lies in the way it inverts the connotations of infrastructure as a mechanical and systematizing extension of authority. When "extende[d] . . . directly to people's activities," infrastructure becomes something populist and alive, "characterized by incessantly flexible, mobile, and provisional intersections of residents that operate without clearly delineated notions of how the city is to be inhabited and used" (Simone 2004b, 407).

Intuitive as this idea is, it raises important questions about how "incessantly flexible" a social field can be. Michael Watts, for example, sees something exhausting in Simone's street-level description of permanent freedom: "It is a sociability built out of 'trickery,' an 'incessant state of awareness,' of protections and payoffs, of duplicity and deceit, of guardedness (not a Simmelian blasé detachment), of operators and scam artists—a world in which 'anything can happen' (and is this not a sort of terror?)" (2005, 184). I have likewise come to wonder if in such accounts the African city doesn't begin to resemble a kind of spot market, where there are only small-scale contingent transactions, and thus where people aren't really persons at all in any meaningful sense, but strangers with temporary common interests. And while this is undoubtedly one level of urban life, are there deeper, perhaps more 'infrastructural,' levels to be found? Are there senses in which, to the contrary, it is important for people to be enduring, stable, and predictable?

With its melding of the animate and inanimate, "people as infrastructure" shares a family resemblance with another concept that has traveled well beyond its Africanist origins (e.g., Rogers 2006; Kusimba 2020): "wealth in people," which generally refers to historical practices of clientage and dependence (Miers and Kopytoff 1979), as well as to broader kinds of self-making through enskillment (Guyer and Belinga 1995). This is a perhaps subconscious connection here that to my knowledge has remained unexamined, but I would suggest that bringing the two together adds a depth and historicity to Simone's rhizomatic formulation and, in the case of *vishoka* at least, suggests a slight but decisive modification: *personhood* as infrastructure.

The literature on wealth in people has long emphasized the distributed and processual nature of becoming a person in African contexts (Fortes 1959; Jackson and Karp 1990; Guyer 1993b; Karp 2002; Fioratta 2015). "With age and time," Sibel Kusimba (2020, 168) summarizes, "men and women moved through the life cycle acquiring rights-in-others and achieving positions of prestige. Women gained wealth-in-people by initiating girls and providing them as brides, by controlling cooking and food in households, or through the children they provided to husbands and lovers." Across the arc of their biographies, in other words, individuals sought to amass various kinds of "capital" (land, children, skills, livestock, currencies, dependents) and assimilate their power in what Ghassan Hage (2013, 79), following Bourdieu, calls a "political economy of being." As in other ethnographic contexts, this vision of personhood as a state of becoming renders the distinction between people and things ontologically permeable (Strathern 1996; Viveiros de Castro 2014; Vokes 2013). Nyamnjoh speaks of "convivial" or "composite beings" (2017, 260) whose agency is "cultivated, nurtured, activated and reactivated to different degrees of potency through relationships with others, things and humans alike" (256). The mutual becoming of people and things thus resonates with Mauss's ([1925] 2016) basic insight into the Gift: that, in its circulation, it channels the spirit of the giver.

In many African societies, however, not all people + thing becomings are equal: some are relatively more person-like while others are more thing-like.[1] An often-defining feature of male youth, for example, is their protean, relatively *pre-social* quality. Possessing neither wife nor children nor property, youth were in effect derivative extensions of their elders who could enroll them in projects of collective labor or violence—a system that McGovern (2010, 56) has called "gerontocratic-hierarchy." Ideally youth become elders in the "organic" process of maturation and decay of the domestic cycle and mediated by ritual performance (Goody 1958), but rarely does this process of succession precisely or evenly unfold, since elders usually profit from juniors maintaining their subordinate position as long as possible (Fortes 1959). In extreme cases the imbalance gives rise to a "crisis of social reproduction," in which even aging men are embarrassingly confined to the social status of youth (e.g., Hickel 2014). Alternatively, subordinates might "jump the queue" and claim social power through different kinds of commerce, war, or other dangerous potencies—an ambiguous agency that McGovern calls entrepreneurial-capture" (2010, 56–57). But, as with all social insurrections, "youth" in positions of power are faced with the "day after" task of legitimizing themselves.[2] These dynamics share a family resemblance with contemporary Tanzanian political idioms of "patronage" versus "rights" outlined in chapter 2. Does power flow from what you've come from (and thus who you are) or what you are doing, here and now (and thus who you are becoming)?[3]

On the precolonial East African coast, *utumwa* (slavery, or servitude) was a key site where such questions about personhood, action, and community were negotiated. The term is a complex one in the regional historiography (Miers and Kopytoff 1979). While born of the violence of the nineteenth-century caravan economy, it is best understood as "one of several coexisting varieties of subordinated client," partly analogous to youth, women, and other social dependents with some degree of autonomy (Glassman 1991, 286).[4] In his influential account of the "cultural biography of things," for example, Igor Kopytoff (1986) described how, once captured, a slave is stripped of all the kinship ties that make him or her a unique individual with a history in a community. Slaves in this raw state were a kind of a liquid labor power, a generic force without form. However, upon entry into his or her new situation, re-formation was possible. He or she could acquire degrees of social personhood through accumulating money to pay for bride wealth, religious conversion, or community rituals. While retaining the ascribed (anti)status of a slave, the cultural tolerance for reducing him or her to a kind of brute energy was reduced. As Hart (2005, para. 18) notes in his discussion of money, "the aristocracy everywhere claims that you cannot buy class. Money and secular power are supposed to be subordinate to inherited position and spiritual leadership." And while elites might look down on such *arrivistes* who take a shortcut to high status, they may well be forced to deal with them, however begrudgingly or discreetly.

Interestingly, what Kiswahili speakers now call wage dayworkers (*vibarua*) and craftsmen (*mafundi*) originally referred to skilled slaves involved in specific kinds of tributary relations in the rapidly commercializing regional coastal economy of the late nineteenth century. These mostly male slaves were entitled to conduct petty commerce in the Swahili cities or on caravan routes, provided some percentage was remitted to their masters. With their earnings, slaves could purchase titles in local dance societies and therefore urban respectability. The more ambitious among them obtained access to credit and organized their own trading operations, sometimes outstripping their masters in commercial prowess or absconding from their authority completely. In short, commercial success could be converted—albeit at uneven exchange rates—to degrees of recognition within one's community, often to the begrudging resentment of their masters/patrons.

Jonathan Glassman's (1995) magisterial history of the late nineteenth-century Swahili coast charts the emergence of a self-conscious Arab planter capitalist class that, combined with German intervention, partially flattened the moral economy of the caravan routes and Swahili city-states. Nevertheless, its idioms and practices that began as far back as the 1850s do persist in the region, particularly around questions of labor. Sarah Hilleweart (2016), for instance, charts the way "Beach Boys" on Lamu Island off the Kenyan coast—poor young men wooing Western tourists—are sometimes denigrated as "modern-day slaves" for the ways they forgo customary, Islamic modes of dress and comportment. Beach Boys, by contrast, emphasize their work as a form of material and social striving, one often discreetly tolerated by their families for the financial contributions it engenders. As in earlier "liberalized" eras of East African life, a stance of "entrepreneurial-capture," an individualized ability to open up access to expanded circuits of money and value, becomes a potent if morally ambiguous means of social transformation.

Analogous dynamics can be found in the world of electricians, contractors, and utility employees in Dar es Salaam. There is often an entrepreneurial impulse among low-level *vibarua* and *mafundi* (*fundi* singular) electricians, who, in conversation with me, sometimes explicitly described selling their raw labor power as a type of slavery. *Vishoka* is a name for those electricians who build their careers by contriving deals that benefit Tanesco bureaucrats, field crews, and inspectors. One of the ironies of the *kishoka* is that he may start out as a Tanesco *kibarua* but, by going into "private practice," can eclipse financially the salaried civil servants on whom he depends for materials and expedited service—a fact that the latter resent. Like the slave on the Swahili coast, a *kishoka*'s ascriptive status does not exhaust his substantive standing in the community. There are certainly deep and important differences between these two figures, particularly the threat of violence to which the former was subjected. Nevertheless, the parallels are noteworthy and, most importantly, were sometimes drawn by my interlocutors themselves. Both figures attempt to combine their craft expertise with investments in social relations to achieve a kind of singularity, such that their reputation and skill name a reliable pathway through which value flows. We can speak of such figures as "becoming infrastructure" in a way that resonates with Simone's evocative analogy, but this is an infrastructure that is both dependent upon and produces a certain kind of socially recognized personhood.

Moreover, there is a recursive pattern here in which *vishoka* not only become infrastructure but become it at deeper and deeper levels. Roughly speaking, they often start out at the front end of the distribution network—the consumer end, or "street"—and ideally work their way to the producer end, or Tanesco office. This entails amassing a track record of successful dealings with both customers and with their patrons inside Tanesco itself. As *vishoka* reputations expand, so too does the scope of their interventions. They move from the relatively superficial act of channeling the flow of electrical current (through minor repairs, wiring or tampering), to the upstream *channeling of those channels*—that is, to the disbursement of the actual socio-material network (meters, wires, poles, and bureaucratic approval). To borrow the language of Nancy Munn ([1976] 1992), we could imagine *vishoka* as operating at a more expansive level of "intersubjective spacetime" by acting as the medium through which the network is not just maintained, but extended. Profits at this level allow them a kind of social authorship, or what Jane Guyer calls a "self-realization" of their own lives (1993b). Indeed, many *vishoka* benchmark their own success in life by another kind of infrastructural platform discussed in the last chapter—the building of a house. As in many other places around the world, the "petty bourgeois" (Scott 2012, 84-100) desires for small property and some economic autonomy are key indexes of proper social reproduction—indexes of "the good" (Robbins 2013).

Still, at whatever point up and down the network where *vishoka* find themselves working, their unofficial presence gives them an essential ambiguity. In "fixing" the grid, *vishoka* are neither state actors, nor are they private citizens. In some ways, they take on the worst qualities of both parties—sometimes seeming to abet the enterprising indiscipline of urban consumers, and sometimes seeming to embody the rent-seeking venality of the state. Are they social *parasites*—private, selfish citizens who are unduly profiting from the failures of a public good (*mali ya umma*)? Or are they social *channels,* a populist force subjecting a de facto luxury good to the lubricant of common coin? The answer is that they are both—in dynamic proportions that are worked out over a succession of individual deals, over the course of individual careers.

living on (and off) the last mile of the network

I can say that a *kishoka* can be observed in two ways. If a problem happens in the house, at night for example, I can call a technician to come and fix the problem—which is to say something totally normal. But in another sense a *kishoka* is someone who . . . [works] in a situation where there is a big emergency in the house, and you call Tanesco to come solve the problem, but you can't wait for them. So it behooves me to get someone to come solve the problem. So, he who solves the problems of Tanesco but is not recognized by Tanesco, we call him a *kishoka*. But that first *kishoka*, he's just a normal person, a normal electrician from the street. You call him, he repairs the switch, he does this and that, he's a regular person. . . . So there are two types here: a *kishoka* does work when Tanesco's phone isn't working, and then there's another *kishoka* who solves little problems inside the house. So I take these two aspects in the meaning of *kishoka*.

In some ways this description, given to me by a resident living in the "Swahili" neighborhood of Buguruni, points to the ambiguity of the term *vishoka* (and to the empirical confusions of fieldwork generally): in this telling, a *kishoka* is both a normal electrician and a rule-breaking, trespassing one. But rather than two synchronic categories, I find it illuminating to think of them as a sequence, a kind of primal scene of transformation, where the street electrician becomes the *kishoka* he already potentially was. He does this by making the jump from repairing "ordinary things" to involving himself in the problems of Tanesco. And that jump begins along the last mile of the network—the service line.

Once state-generated current passes through the service line, it diffuses out to any number of physical media: wiring, refrigerators, batteries, televisions, light bulbs. During interviews, many of my interlocutors highlighted that they became interested in electricity through childhood encounters with these objects. One electrician recalled:

> I didn't play football, I liked to close myself up in the house. My grandmother was selling *gongo* [moonshine] and would hide it on the roof . . . if it finished [she would say] "go climb up there and get some more" because I was so small. And that's where I first learned to respect electricity, because I had to jump over the live wires. Then I would think "ah, this one wire passes into the house, where does it go?" Later I came to understand "ok, this is a live wire and this is a neutral one."

Another, now a successful contractor in his forties, offers a similar story.

> I won't leave things until I understand them. I sat for a month, I remember [as a boy] I stole two shillings and went to buy a bicycle light, two wires and a battery, hooked them up, and walked around at night with it. But the thing that gave me trouble was that if I wanted to shut it off I had to unplug the wire. How did I solve that? An idea rushed in: a switch. I found a switch. . . . So it went on like this. Later I came to get a hold of a motor of a tape recorder radio. I removed the motor, I hooked it up to the wires—it worked. I cut up some plastic and attached it to the front, and it became like a fan.

Such is the tinkering ingenuity of Africa's "repair worlds" (Jackson, Pompe, and Krieshok 2012; De Laet and Mol 2000). Discussing discourses of waste and recycling in colonial and postcolonial Dar es Salaam, Brownell observes: "if the urban poor were polluting vectors of disease and wasted rural labor forces, they were also the stewards of objects, developing ingenious methods of repair or reinvention, extending the life of stuff or creating new objects entirely. . . . Navigating through the strange new territory of both consumer objects and urban poverty, townsmen . . . created charcoal stoves from car doors, lamps from oil tins, and tambourines from bottles" (2014, 223).

Much critical theory has found something almost utopian in this stewardship. Jackson (2014, 237–38) invokes Walter Benjamin as the "patron saint" of broken-world thinking, who, amidst the debris of history "quietly goes about the business of collecting and recuperating the world around him." Notable here is the way Benjamin himself associates that recuperation with children, who "do not so much imitate the works of adults as bring together, in the artifact produced in play, materials of widely different kinds in a new intuitive relationship" (quoted in Buck-Morss 1991, 262).[5] Indeed, for many electricians I spoke with, their early encounters with electricity took place in a postcolonial environment that, resonant with childhood itself, was both marginal and mutable—a gathering of virtual connections that a clever-enough tinkerer might actualize and sustain.

This kind of environment was also important for the way it could set youth along the path to becoming electricians, sometimes accidentally. A contractor named Besti, for example, recalls how he initially went on to postsecondary education for a different specialty entirely, but found he was getting more work as an electrician and decided to continue with it instead. On the other hand, formal electrical training is often a remedial career path. Many electricians describe how they were unable to pass their Form Four exams that would allow them to continue to higher education and decided to enroll in VETA, Tanzanian's vocational schooling system. After certification, there is a wide range of trajectories. Some manage to find work as electricians in banks, hotels, or other industrial or commercial enterprises, but poor pay and difficult working conditions might push them to combine this with any freelance repair work or temporary contract work on construction sites. Others still might supplement or substitute these arrangements with so-called *vishoka* work, which may include a range of different activities.

When I told a friend in Dar es Salaam about my fieldwork with electricians, he chuckled and suggested I meet Issa, a man best known locally for injuring his leg while jumping down from a utility pole in an attempt to run away from the police. He was, the friend said, grinning, "a real *kishoka*." When we finally did sit down at a local bar, Issa did not disappoint. He arrived theatrically, announcing that he was tired of getting shocked—the last one was enough to swear him off utility poles altogether. Working on a transformer, he had had the misfortune to grip his pliers "poorly," where there was no rubber insulation; after that, he said laughing, "only God himself knows what happened to me; all I know is I came to and suddenly had to pee." Minor shocks are inevitable, he went on to reflect, the mark of any electrician whose work entails *mambo ya practical*, the practical problems of the handyman. He memorably likened the jolt of current to a soccer ball shooting off the cleat, and his own constant movements around town to that of a skittish crow, flying from branch to branch.

In fact, during our conversation, Issa stopped in the middle of his animated explanations to answer his phone: a woman wanted to know if he could fix a short. "I'm a bit far," he said, "let me hop on a motorcycle taxi and I'll be right there." Actually, he smirked as he hung up, the woman was right down the street, but this way he could charge extra for *nauli*, travel fare. Without much ado, he jumped up and walked away, leaving his beer and half-eaten chicken on the plate. Fifteen minutes later he came back, having decided not to do the work. The household's other tenants were out and wouldn't be returning until morning, and the most money the woman could muster was Tsh. 4,000. And besides, he was getting drunk and in no mood to get shocked again. He would return later.

This hustle in the sense of both exertion and trickery (cf. Spence 2015) was a common feature of such "street electricians" (*mafundi ya mtaani*) consigned to continuous improvisation, up to and including running off in the middle of dinner. In part this was due to the casual nature of the work and its narrow margins. Another street electrician named Morgan described how his customers tended to underpay him, balking at a price if he should finish a job especially quickly. "Tsh. 10,000 just for that? It took two minutes to fix!" What they failed (or feigned failure) to see, he explained ruefully, is that "the problem isn't working on the problem; the problem is finding the problem." In other words, they framed the work as the mechanical transfer of raw labor time to the object, alienable from Morgan's particular craftsmanship and diagnostic skill.

Issa, however, did not simply rely on such small-scale repair work. Over the past decade, he had also worked on and off as a Tanesco *kibarua* in construction and service line installation. *Vibarua* work, as we saw in the last chapter, is generally resented for its low pay, difficult work, and lack of long-term security. In this way, working as a *kibarua* was not much different than street work. One the other hand, it also provides opportunities to build up skill and leverage social connections. As Besti observed:

> You could come across someone who has worked for Tanesco for ten years, but on renewable three-month contracts, and that's their livelihood. All of a sudden, ha—that contract comes to an end? What do you expect him to do? Firstly, he knows many customers, because he's long been doing such work, and secondly, he knows the whole setup in Tanesco [*setup nzima pale Tanesco*]. He and his boss can have deals together—and it's not just they

know each other in passing. They've come to trust each other. And so he has no problems calling him for something on the side—there is some kind of mutual understanding among them.

Indeed, Issa's time in Tanesco has given him skills not taught at VETA. One of the ways he now supplements his income is by helping residents with the problem of low voltage. Peak demand, the general increase of new connections to the grid, unofficial line sharing, and poorly insulated or dirty wiring can all contribute to low voltage. In addition, wealthier households or businesses with a three-phase connection suffer load imbalances, in which the alternating currents that flow through the line leading out from the transformer fall out of sync. Refrigerators may run but never quite get cold; mobile phones may take an entire day to charge, and in general, electrified life grows slower and weaker. For around Tsh. 15,000, Issa will trespass up onto the transformers and rebalance the load—a job that Tanesco is supposed to do for free but rarely does. It is in this particular sense that he is a *kishoka*.

Some wealthier residents have hired Issa to tamper with their meters so that the kilowatt hours are recorded more slowly. The safest way to oblige this request, he explained, is to distribute the risk through collaboration. Issa says he might charge the customer Tsh. 250,000, pocketing 50,000 upfront and passing the rest along to a Tanesco foreman. Issa will be sure to keep an eye out while his partner works, but his green Tanesco uniform and company truck parked outside provide most of the camouflage. A Tanesco employee likely has easier access to ladders, protective gear, and replacement plastic seals, whose breach designates the meter tampering.

As we will explore below, this collaboration may hinge on trust and mutual understanding, but the relationship between foreman and temporary workers is not free of class conflict. In interviews, *vibarua* resented permanent Tanesco employees' air of superiority, and derided them as nepotistic hires with little practical experience. After all, Issa told me, "the slave doesn't like to be ordered around" (*mtumwa hapendi kutumwa*). Still, such moonlighting helps soften the edges of Tanesco's employment hierarchy, and allows it to continue—albeit not without tensions—even after its formal dissolution. Issa does not presently have a contract with Tanesco, but the foreman-*kibarua* dynamic remains.

Issa's forays into load balancing and meter tampering illustrate a kind of zero-level case of becoming infrastructure along the last mile of the distribution network. Michel Serres's central insight in *The Parasite* (1982) is that no relation is reducible to the relata it connects. Even a perfectly functioning "black-box" takes its cut, inherently "sieving" out some qualities of its input to create an output (Kockelman 2017). Think of the difference between the same message delivered by text or voice, or (see introduction) the difference between stirring with a teaspoon versus digging with one. Any given channel is one mode among a number of possibilities that shapes the character of a given act and the relation it creates (and it is for this reason that all channels are, paradoxically, also parasites—they never give you the "the thing in itself"). Still, while this is true in a general sense, it is also true that in any social context, a given channel is more or less predictable, and hence is a more or less faithful expression of its relata. Building on Serres, Bruno Latour marked this as the difference between intermediaries and mediators (2005, 37–40), and both, as Kockelman points out, echo Charles Pierce's distinction between secondness and thirdness: "By the third, I mean the medium or connecting bond between the absolute first and last. The beginning is first, the end second, the middle third. The end is second, the means third. The thread of life is a third; the fate that snips it, its second. A fork in the road is a third, it supposes three ways; a straight road, considered merely as a connection between two places is second, but so far as it implies passing through intermediate places it is third" (quoted in Kockelman 2017, 37).

As handymen (*mafundi*) or dayworkers (*vibarua*), Issa and Morgan are, relatively speaking, intermediaries, or seconds. Their minor repairs of wires and appliances create short-term lateral equivalencies between cash, current, and commodities. They are always available, ready to do this job or that job. They are a bit like what Hannah Arendt (2003, 167–81) describes as anonymous "labor," a metabolic cycle of equivalent inputs and outputs, decay and repair. In aggregate their labor is useful for maintaining the flow of electricity at the street level, but no particular "beat" is itself very significant.

Against this background, it is fitting that Issa would describe *kibarua* employment in its original sense of *utumwa*, slavery. On the spectrum of clientelist servitude, a slave was at the far end, the most derivative, indeed most commoditized, extension of a master's singular will. The verb form of *mtumwa* literally means to be sent, a condition of bonded motion, a turbine spinning in place: constantly moving but never getting anywhere. Likewise, a Tanesco *kibarua* is a subordinate laborer who is dispensed out to the work site, and whose value is dispensable, worth only as much as his time on any given day.[6]

As an employee Issa may be an intermediary between Tanesco and citizen, but as a *kishoka* he is a mediator. He hooks the dendritic edges of the state grid into commercial circuits of repair and modification. This has a double effect. First, it conceptually nudges state-owned meters and transformers closer to refrigerators, light bulbs, or radios—that is, it partially downgrades them to relatively commonplace commodity objects amenable to tinkering and market exchange. In turn, the act upgrades Issa, making him more concrete and socially distinct. In these situations, he is not just an interchangeable tinkerer with only his raw labor power to sell. He is someone in particular, someone that people that would find good to know. By expanding his spatio-temporal influence over the circulation of electricity, he diminishes his own socioeconomic marginality.

It is worth attending to the way the affordances of the network allow for Issa's becoming infrastructure. Because electricity will flow to whatever conducts it, sociolegal distinctions concerning its pathways are blurred in the phenomenology of neighborhood life. A young kid inclined to climb the roofs and follow the pathways of electrical current will find wires, radios, and fuse boxes strewn about his world. Should he pursue work as a street electrician, he will find that his entry-level repair jobs, where electricity is objectified in private commodities, lie right on the border of Tanesco infrastructure, where it is still state property. He may find that it's a short from jump from (relative) secondness to thirdness—from repairing a circuit inside the house to mediating a service line right outside of it.

routes, actual and possible

Is *kishoka* work legitimate? Yes and no. According to one resident living in Buguruni, an unplanned settlement, *vishoka* thrive because

> Tanesco doesn't look after its own work. They don't care about citizens' time, because when citizens come to report a problem with the power on their street, if you call on them today, they come after ten days. And back at the shop or the house, what are you going to do if there's something dangerous—what can you do, brother? You have to find a shortcut. You say, "Ah, I'll call up a *kishoka*, give Tsh. 5,000 or Tsh. 10,000, the issue will be resolved easily." So Tanesco has to improve their services in order for *vishoka* to leave, in order to restore the citizens' belief in them.

What kind of reasoning is this? To illustrate modal reasoning, Verran (2007, 171) offers the example of ranking things on an ordinal scale such as first, second, third. Like adverbs, ordinal rank describes an overall relational pattern. It alone does not tell us anything about how much and of what quality the first of a given set is "more" than the second or third; it tells us only that they share, at minimum, some general relation to each other.

The unofficial repairs that prop up the network in places like Buguruni work the same way. Calling *vishoka* hardly ranks as the "first choice" for residents or the utility. They contravene Tanesco policy by trespassing onto the street edge of the distribution network, and for Tanesco this cannot but evoke the specter of vandalism or even theft. Its website, brochures, and other public relations statements warn that *vishoka* services are illegal and that they are poorly educated amateurs who might do more harm than good. For their part, residents do not seem much happier to use them. If one is forced to call a *kishoka*, one is always already paying extra, and this extra can spiral to truly taxing proportions.

Vishoka repairs are thus a "second choice" for residents, but insofar as they are a second choice, they share the general family resemblance of all viable choices in the set of possibilities. They restore technical functionality at the edges of the network and thus, retroactively, Tanesco's monopoly on it (since that monopoly is itself reciprocally predicated on technical functionality). They eventually realign the edges of the network back into a feedback loop, and this makes them different from official repairs only in degree, not kind. Hence residents understand that while *actually having* to pay *vishoka* amounts to a risky or burdensome proposition, it is legitimate to be *able* to pay those same *vishoka* when Tanesco is not responding. Whether or not *kishoka* work is actualized, it should nevertheless be preserved as possible.

Despite their warnings, Tanesco's managers also have good reasons for tolerating *vishoka* repair work. A decade of austerity has institutionalized a labor force of temporary contract workers and underpaid salaried employees. Tanesco can partially outsource its labor and administration costs by letting them "pay themselves." Ceding a low-level autonomy to *vishoka* at the outer edges of the network may be worrying, but Tanesco has little incentive and capacity to dissuade or prosecute them for making repairs, at least relative to other unofficial activities such as such as theft or tampering, which amount to a species of vandalism.

Things get trickier when *vishoka* venture past merely repairing service lines to tampering with them, though here too there is a moral gradient that runs from the more tolerable to the less tolerable. At the former end, there is the phatic act of reconnection through the meter. Like repair, a reinserted cutout is oriented to nothing more—and nothing less—than restoring the channel that links utility and consumer. At the other end, outright theft severs the system's logic of reciprocity. But even here, certain distinctions can be drawn. In contrast to the bypasses explored in the last chapter, Issa prefers to finesse his tampering jobs by routing them through a Tanesco patron. The logic is this: to tamper is to find oneself trespassing onto salaried Tanesco workers' turf, and thus any would-be thief would do well to render some of the profit in exchange for their patronage and protection. A bypass may well constitute stealing from Tanesco, but insofar as Tanesco's own employees consider themselves overworked and underpaid, their involvement can function as reclaimed income. We can also note that *vibarua* or their Tanesco patrons will likely have the knowledge to contrive more sophisticated bypasses—in this case by simply slowing the digital recording mechanism.

Just as a *kishoka* can fully side with the consumer at the expense of Tanesco—by, for example, straight up bypassing a meter, he can also fully side with Tanesco at the expense of the consumer. A resident once described to me one such scenario wherein a *kishoka* bypasses a household, then turns around and reports that household in order to extort a bribe on the back end.

> For example, a *kishoka* can come by. This just as a little example here, this kind of thing is common in the Swahili areas because people have long tired of Tanesco's narrow service and so there are lots of scams. He will tell you he can connect you illegally because [you expect] Tanesco to bring service. If you've been robbed of your meter, for example, Tanesco is supposed to come and return your service. But Tanesco doesn't come, so you'll go and see a *kishoka*. He comes and does a little connecting in some way of his own devising and your power returns. But later that *kishoka* who's just connected you illegally—the very same one!—he'll go to Tanesco real and say "this particular house, it's stealing power." And then Tanesco real they come. So they've arrived and what happens? They'll demand a large ransom. They'll threaten to send you to the court because you've sabotaged the company. You can't accept the scenario of going to court because it's a huge amount in fines for sabotage. What will you do? You'll sit and talk with the guy and give him a little something. You'll issue Tsh. 500,000 then and there. Tsh. 200,000 goes to the *kishoka*, and Tsh. 300,000 goes to Tanesco themselves. You see? So a *kishoka* plays two scams. He does some work for you and then turns on you by going to Tanesco.

As these case studies show, there are many possible variations when *vishoka* meddle with service lines. Depending on the nature of their intervention (repair, reconnection, bypass) and the degree to which Tanesco employees are passively or actively involved, *vishoka* can restore or degrade the long-term relation embodied in the service line, sometimes even doing both in a complex sort of sociotechnical origami. The main point is that the presence of *vishoka* on the service line pushes the relation between consumer and utility closer to the living dynamism of thirdness. There is still a relation between input and output (current and currency), but the "doing" of that relation, the actual work of conversion and transduction, is no longer black boxed. To the contrary, it is foregrounded and thematized as a negotiation with its attendant "marginal gains" and losses (Guyer 2004). The presence of *vishoka* thus rouses the service line from being merely actual—a route to a destination—to a possibility space, a logic of possible routes (Kockelman 2017, 74–76). It is in this sense of experimentally incorporating themselves into the channel that links consumer and utility, of seeing which ways it might bend and shift, that we might think of electricians like Issa as "becoming infrastructure." And just as *vishoka* seek to "jump up onto" (often literally) the decentralized edges of the network, they also seek to position themselves further upstream, along its institutional trunk. That is, they seek to mediate the process of extending service lines in the first place.

living on (and off) future lines

After years of piecemeal work, an *mzee* (elderly man) was finally nearing completion of his house in the well-to-do Dar es Salaam suburb of Mbezi Beach. Within the house's cavities, neat bundles of wires threaded through networks of blue plastic tubing. A contractor had diagrammed their pathways on thick cardboard paper, stamped the bottom corner with his company license, and submitted an application to the local Tanesco branch office. Two weeks later, a Tanesco surveyor was to come out to Mbezi Beach and approve the structure. The *mzee* would pay a Tsh. 750,000 installation fee at the office, whereupon an installation crew would mount a bracket near the roof, drill a meter to the side of the house, and link it to a secondary cable drawn from the nearest utility pole. But the date had come and gone, and the surveyor had never shown. The *mzee* was anxious, especially since he was to be away, traveling outside the city for a funeral.

Unsure of what to do, the *mzee* asked a neighbor for advice and received the phone number of Samueli, a former Tanesco dayworker. Samueli took a minibus out to Mbezi that very evening and appraised the situation. "I told him the truth," Samueli said to me. "I said, 'Don't go in empty-handed. They will just hassle you.'" He offered to return the next day with the surveyor for Tsh. 100,000. The *mzee* produced Tsh. 80,000, withholding the rest, and said, "Until Tanesco arrives." Samueli agreed.

So I called up my surveyor and told him he had some drinking water (*maji ya kunywa*). He said, "fine let's see each other tomorrow." We met up the next day. I took the Tsh. 80,000 with me, met the surveyor and gave him Tsh. 30,000. We took a taxi over, and I paid for the taxi, Tsh. 10,000. We came over and the surveyor measured everything. I told the *mzee* to wait inside and just let us work. The surveyor finished and left—no problem. I went inside and the *mzee* gave me the remaining money. I put that together with what I had and saw Tsh. 60,000. Great; I went home and bought flour and rice for my family, fresh. My family eats. And that's the style in which I live in the city.

For the motivated and/or talented electrician, there is good money in wiring the homes and buildings that have mushroomed across the gentrifying core and peri-urban satellites of Dar es Salaam. Consumption of electricity by the city's light bulbs, stereos, and TVs may wax and wane over the course of the day, but these electric cilia depend on wired residential and commercial structures to feed them current. Unlike the scrum of spot repairs or temporary wage labor, wiring up a building thus constitutes a higher order, more encompassing level of the private side of the network. This fact is expressed in its economic value; wiring involves multiple kinds of parts and labor and thus multiple margins across which to calibrate gains and losses. But wired buildings do not just feed downward, they are themselves fed by something higher still: the public network. The whole fractal unfolding of the power "feed" cannot happen unless this primary connection to Tanesco, and the long-term social commitments it embodies, is secured.

Unfortunately, one of the major frustrations for the Tanesco consumer is the very process of becoming one. Thanks to decades of disinvestment as well as NetGroup's more recent austerity programs, applying for a new service line is subject to all manner of shortages, delays, and interferences. The number of applicants overwhelms Tanesco offices, and the process can stretch out from three months to two years. As they do for existing physical service lines, *vishoka* can, for a price, bridge Tanesco's institutional recalcitrance, helping to midwife new service lines into being and in turn further expand their own careers.

How then did Samueli simply call up "his" surveyor? After graduating VETA, Samueli lucked into a job through his neighbor, a Tanesco employee who knew his skill as a street electrician. Working as a service line construction crew dayworker (*kibarua*), Samueli was paid Tsh. 5,000 per day to install brackets and meters in houses that had completed the service line application process. He supplemented this meager paycheck, which often came late, with "drinking water" (additional revenue) that the foreman would distribute to the crew after unofficial jobs. After four years, however, Samueli was part of a cohort of eighty *vibarua* whose contracts were not renewed—the official reason being that they had not passed their Form Four educational exams. When I asked Samueli how he felt when he learned he had been fired, he said:

> Ah, when I was fired I felt bad. I said, "truly, the old man who gave me the job didn't have any pull [to keep me on]." So I said "now I'm in the street." I was even happy to leave . . . well you know, I wasn't happy about it but I already knew how to be a technician, I knew lots of things well. [When I started] I didn't even know how to install a meter, I didn't know how to differentiate between 33kv or 11kv, or know LV [low voltage] or transformers, but I learned all that in the company. So when I left the company it was sad, [but] I thought "better I do this street work that goes by the name of *kishoka*." I saw that I should do this kind of work to move my life forward.

Cut off from the official educational and employment ladder, Samueli reconstituted his career "in the street." And by acting at the hatchet in the *mzee's* otherwise empty hand—that is, by channeling his money to the surveyor—Samueli was able to expedite access to a new service line.

The commercialization of speed (and bureaucratic discretion, as we will see below) creates a range of service options that reflects a local class gradient while maintaining a baseline of common access. As Besti put it, "A poor customer may want things to go faster, but because of his limited resources, he will just decide to sit there on the ground [*kukaa chini*, i.e., wait]." By contrast, wealthy residents may pay their electrical contractor to act as a *kishoka* after the wiring is done, piloting the entire service line to completion. In such full-service deals, the *kishoka* ensures the "content" of the procedural form, the human touch that translates the rule into action (Hart 2006). The effect is something like tiered service on a commercial airline. A premium ensures a smooth experience in the form of priority boarding, cancellation insurance, and an attentive, quasi-personalized staff.

There is also, however, a midrange between the extremes of the poor "paying" in time or hassle and the rich simply throwing cash at the problem. In some combination of ignorance, optimism, or economy, customers may opt for the official price and hope for the best. If an obstacle comes up, they might turn to a *kishoka* for help with a specific issue, as did the *mzee* with Samueli. Hence, *vishoka* are "modal add-ons" (Verran 2007, 173) to the service line procedure. They are not essential to any particular step, but rather reflexively mediate the pattern of steps as a whole. Note, for example, that they do not charge for the actual service (the surveying, the disbursement) but rather, more ambiguously, for the *conditions of access* to that service. This subtle distinction is important because it preserves Tanesco's nominal monopoly over the service line application procedure. The procedure can shift modes: it can go slower or less slow, or it can be expensive or less expensive, but it must ultimately be secured by Tanesco; otherwise even the basic pretense of reciprocity animating the relation becomes untenable. We can elaborate this point in the following section, concerning the trickier question of altering access to the very materials of the service line.

the *kijana* and the mama

Over a few drinks at a local bar, Besti relayed a recent job he had acquired from a certain *Mama*, an elderly woman of around sixty, living in an inherited house (*nyumba ya urithi*) in the Karakata district of the city. Besti was to help her apply for a new service line—or, rather, reapply. Two years ago, she explained, she had unexpectedly come into some money and decided to "get *umeme*" for her household. She hired an electrician to wire her house and asked for his advice on how to get connected. "There are people that can take care of this issue," he told her. "I'll give you the number of one, and he can handle it."

A few days later, a *kijana* (youth) arrived at her home. As they got to talking, the *kijana* explained that she had two options. She could take the "official route" (*njia rasmi*), certainly, but that could cost her time (*itakugharimu muda mrefu sana*). She might wait up to a year for electricity. But there was also an alternative route (*njia mbadala*). "And the Mama said to herself 'I came to this money by the grace of God, by luck. If it will take a long time, I'll just fritter it away.' And so she opted for the latter."

Here the Mama faced a classic problem of Swahili life, and indeed of many distributive economies in African contexts. Because the livelihoods of urban dwellers often depend on value continuously circulating through social networks (as credit, debt, tax, tolls, and so on), saving and accumulation are very difficult unless that value can be amassed and converted "upward" into a long-term asset (Guyer 1997; Ferguson 1985)—in this case, an infrastructural connection. If the conversion process takes too long, one risks proleptic liquidation—losing both the funds and the asset itself.

It's not entirely clear when the Mama actually decided to take the "alternative route." As she explained it to Besti, the *kijana* started more or less "officially"; albeit, in retrospect, questionably so. He took Tsh. 10,000 for the service line application form (which, as Besti noted, only costs Tsh. 6,000). He returned to her house, diagrammed its wiring, and solicited another Tsh. 10,000 to get it officially stamped by a licensed contractor company (this too was overcharging). He told her he would have to return for the surveyor, and the Mama waited for about a week.

"At this point in the story," Besti narrated, "I started to get concerned. I had a hunch that he was on a mission looking for a meter." And indeed, when the *kijana* returned, he told her he had yet to secure the surveyor's visit. But it so happened he had a meter, so why didn't he just go ahead and install it? The meter was registered under a different name and to a different ward of the city. But not to worry—she could eventually get that changed, In the meantime, she'd be able to use *umeme* like normal. If she could provide him cash up front, the whole "exercise" (*zoezi*) could be done in two days. The Mama agreed and supplied Tsh. 500,000.

That very evening the *kijana* returned. He affixed a bracket to the side of the house, secured the meter, and placed plastic seals over the casement. The next day he returned in a Tanesco truck with a few more men, who extended a cable from the nearest utility pole to the bracket. "In other words," Besti said, smiling, "after 15 minutes they left with the power all connected and the Mama with electricity." She was given a receipt with the meter's registration number, which she would use to buy tokens.

This arrangement remained in place for two years, until a DC patrol passed through her neighborhood and inspected her meter. When the Mama could neither explain why, upon providing a receipt for electricity credits, the meter was registered to a different area, nor produce its registration card, they informed her that it seemed to be stolen, and (to her surprise) simply instructed her to report to the nearest Tanesco branch office. Two days later, when she went to a vendor's kiosk to buy more credit, she was informed her meter was blocked. But she was still afraid to go to the branch office; her neighbor warned that she would be exposing herself to legal action, and suggested she get in touch with Besti and ask his advice. Besti considered, and then replied.

> I told her to go straight to the office manager—but you have to lie a little bit and you'll have to take the loss. But better to lose service than get locked up. Say: "yes it's my house, but I was renting it out. I wasn't there, I had gone out to the countryside [*mikoani*, i.e., the village]. When I returned I found my tenants had all chipped in to connect the power. And when the inspectors came the other day they said the meter isn't theirs, well, how would I know? I wasn't around."

Here Besti was advising a bit of *ujanja*; as an elderly woman it would be easier to plead a bit of ignorance and expect a bit of deference. And as a property-owning elder woman it would be quite plausible that, while staying in her natal village or region, her relation to her tenants would be provisional and itself vulnerable to *ujanja*. The ploy worked, or at least did not further incriminate her; the manager told her that a team would return to her house with her and remove the stolen materials, and she would have to reapply—the job with which Besti was now entrusted.

In some ways this story hews closely to a popular genre of urban tales in which greedy or naïve urbanites who tarry with clever *matapeli* (conmen) get what they deserve (Lewinson 2003, 17). But was the Mama greedy? In her telling, it was the worry that her windfall might get frittered away on other distributive claims (from kin, friends, other expenses) that prompted her to move relatively fast, especially in the case of Tanesco's supposed recalcitrance. The motivation here is closer to a hedged bet.

Was the Mama, then, naïve? During the course of their conversation Besti asked the Mama why she didn't follow up with the *kijana* to get the meter registered in her name. After all, had she done so she might have avoided getting caught. She claimed that she had tried to but after a few weeks his phone number was unavailable. And this brings us to the crux of the ways in which this particular *kijana* had pushed the dynamics of modification too far.

> We call these sorts of guys *vishoka*, quacks. Basically, he doesn't care about the repercussions of what he does. He only cares insofar as he gets what he needs. Like neglecting to provide his contact information like that; if a problem comes up how is it going to go? You know, it was perfectly possible to actually change the location, to change the name and none of that would have happened! You know? But he didn't hustle [*hakuhangaika*] on any of that. He saw that it would cost him money.

Besti is, by his own reckoning, not a *kishoka*—a figure defined in his mind by a certain ethical negligence. But Besti is not saying that the shortcuts a *kishoka* might provide are categorically wrong. While there are many alternative ways to arrange a connection to the grid, one should still hew in some minimal way to the social logic linking consumer and utility. We have already seen how component materials bear on this question at the level of *existing* service lines. A service line and indeed the entire distribution network is in effect a social form—an arrangement of material processes governed and animated by a particular purpose, a logic. But its living lines can be severed, reduced to a collection of mere technical objects that can circulate through the market.

Thieves might trespass up onto the network to siphon out transformer oil, or strip a line for its copper wire. Street level *vishoka* also trespass onto state property, but they merely modify the arrangement of materials: for example, by repairing transformers, or confining themselves to reconnections or partial bypasses. Diverted materials from the Tanesco store (seals, brackets, wires, poles, etc.) push this possibility up a level since they do not comprise the disarticulation of existing service lines but rather of *future* service lines. The question is: how can future service line components be diverted in a partial way such that they remain oriented toward the overall goal of growing the network? To what degree can or should one cleave technical objects from the social logics animating them? For Besti, this sort of modal reasoning is what distinguishes a proper contractor like himself from a "quack" *kishoka*.

wire, pole, meter

If we do apply such reasoning, we see that the technical components of the service line can be reflexively sorted into three nested levels spanning the relatively technical to the relatively social—or, in other words, spanning their "capacity to accumulate a history" (Graeber 2001, 34). The first level comprises what we might call "maximally technical" technical components: commonplace, anonymous objects such as wire, cutouts, and seals. These objects are useful for unlicensed repairs of already-existing service lines at the decentralized, street-level edges of the network, or in the case of cutouts and seals specifically, surreptitious reconnections and tampering. They are cheap, generic, and fungible, and as such do not inflict heavy losses when diverted to the street. They also anchor low stakes and for that reason relatively informal and easygoing relationships between Tanesco patrons and *vishoka* clients. Clerks working in the Tanesco storehouse are reportedly openhanded with wire, cutouts, and seals, selling it to their *vishoka* on credit (*mali kauli*; see Ogawa 2006). Dunky, along with an ex-Tanesco foreman named Elson, explain:

DUNKY: If you're outside [Tanesco] you can find someone inside. He knows "ah Michael, he has no problem. What's up?" [You talk about] this and that. "Go [take this] you'll bring it back to me." He trusts you'll bring it back to him, and indeed you will have to. Because you know that problems can happen any day. In other words, if you have a problem today, and someone helps you out you shouldn't say "I won't need help again." There will be another day when you'll need to ask for help. So even if you don't have anything in your pockets, if there's something you need from inside, you can go there and talk to someone, and he'll give it to you. He trusts you'll give him later. And when you go and do that work and get money, you'll give it to him, so that tomorrow he'll give you something again.

ELSON: For example: [you might need] cutouts that are placed on the circuit breaker. Maybe you need wire out at the site. You come to me [i.e. the store employee] and I give it you, 20 or 30 meters worth, saying "*bwana* [mister or friend], go and bring me my money later." Fine, so you go, you sit for a week, two weeks, and then you bring me the money.

DUNKY: And it's not that he'll bother you with any "*bwana*—where's my money?" He knows that you haven't gotten any money yet. And when you get it, you'll flash [i.e., call] him. So it's a unified relationship [*mahusiano ni kitu kimoja*].

Poles fall into a kind of middle level; they are "relatively social" technical objects. Compared to wire, they are riskier to divert and commercialize. They are larger, more unwieldy, and, in recent years, individually numbered. Utility poles, in other words, are not anonymous. Tanesco records their "biographies"—which teams installed them, where, and when. And because they are relatively social elements of the service line, they cannot be completely alienated from the Tanesco store. However, they can be *partially pilfered* insofar as they remain tethered to a new and authorized service line procedure. Here is Besti explaining how this might unfold:

There is an aspect of proximity. Tanesco has its own policy that the wire span from pole to bracket cannot exceed 30 meters. Perhaps [the customer's] house is 60 meters away [from the nearest pole]. A surveyor will come to the house and say that this area needs another pole. Who bears the cost? The customer. And he finds that it is over Tsh. 1,000,000 for one pole! It's too much for him. For one thing he hasn't prepared for it. And for another, a large majority of customers have no idea about Tanesco's procedures or policies. So when this issue pops up, the customer thinks [of the principle of it:] "truly I'm a stranger [*mimi ni mgeni kweli kweli*]! Tanesco, you guys want me to buy a pole? When I'm your customer and the whole idea is to give me service?" And he will find that the answer is: yes, that's the way it is. So the customer will ask, "isn't there some alternative route [*hamna njia nyingine mbadala*]?" And yes those routes are there! And it's Tanesco themselves that lay them out [*wanajitoa mwenyewe*]. I could find someone in a department, maybe Service Line, or Construction, and explain to him the situation. And he'll say, "it's already been entered as a CWO [capital works order]. But I'll tell you what, instead of Tsh. 1,000,000, just get me 500,000, and I'll come and install a pole."

And because I am an experienced *fundi*, I've already analyzed how this situation will play out [*utaratibu huo ukishagundua*] from the beginning, unlike the customer. If the surveyor comes and says the area needs another pole, there's nothing the customer can do if he enters it into the system. But because of my experience I can talk with the surveyor and we can come to an agreement. I'll tell him "Don't write it down just yet, hold on let me devise some cleverness [*wacha nitafute ujanja*], let me see someone in Construction, and have him put down a pole. Then you come and do your survey." So he'll give me time. From there I'll go to a Construction guy and give him this work. Instead of following his lead, I'll draw him into bargaining. If he says Tsh. 500,000 I'll say I only have Tsh. 300,000, [and he'll agree] because there's no loss if it comes from the office [*kila kitu cha ofisini haina hasara*]. He arrives, he installs it, and he puts the Tsh. 300,000 in his pocket. When the surveyor comes, he'll say "for the value of my role in this I'd like Tsh. 200,000 [*thamani ya kwangu nataka laki mbili*]," because now he knows that [the pole installation] has happened outside the office. [He'll say] "In order for me to place [the pole] back in the office, give me 200,000 and I'll push it through."

In short, poles are priced so high as to be antisocial. Indeed, for many residents, it is dubious that they should be priced at all, and thus many feel "truly a stranger" to Tanesco. Through *vishoka*, they can acquire the poles at a discount and thereby qualify for a service line. In this scenario, Tanesco employees still retain some nominal control over the subsequent direction of pilfered materials; it is still a Service Line or Construction department foreman who pilots the installation of the pole, and the surveyor who, on his second visit to the site, looks the other way. In other words, even though the pole was sold illegally, it gets laundered back into the official service line procedure. Or, as a surveyor might put it, the pole gets "placed back in the office," and thereby remains subordinate to the overall goal of providing the consumer access to the transaction—just at a discounted rate.

Finally, whereas the theft and sale of poles can be laundered through official Tanesco procedure, my interlocutors suggested that it is difficult to pilfer meters and that these days they are policed very carefully. Here is Dunky explaining why.

Normally there are things you can complicate; but not those of the community [*umma*]. There are things you can use for your own purposes; but not things like that. Even a thief is supposed to just steal a few hundred shillings. An amount such that you know "he's stealing from me, but it's just Tsh. 150." But he who steals all the way up to Tsh. 1,000 or Tsh. 2,000 . . . it means the next day he'll want to steal Tsh. 5,000.

For example, Tanesco wire. It is sold [by Tanesco store workers] for Tsh. 5,000 per meter. You might find someone who tells you "*bwana* my house has been robbed. Buy some wire for me?" [So you tell him] Tsh. 7000 per meter, and he agrees. So you come and ask Tanesco, they give it to you, you get money, and the days pass. But you know this is just small stuff. And you want something big. Because the last step for electricity is the meter. You'll play around, but at the end of the day you'll need a meter. But a small-fry should leave the meter alone [*hii meter kwa huyu hali ya chini, achananayo*]. You should do small things, so that even if the boss hears "he did what? He stole wire" [it will be ok]. But if he hears you've stolen a meter, you've exceeded. "If he can steal a meter it means he can steal our company's inner secrets" [*siri za ndani*].

For example, a wire necessary for a service line won't surpass Tsh. 100,000. But a meter is more than 400,000. Better to lose 100,000 worth of wire than a meter worth Tsh. 400,000. If you've pushed to that point, you've caused a big loss [to the company]. Even at the shops at Kariakoo, bosses know that "my employee steals. But he won't steal from me to the point that my business dies. He steals 1,000 and buys some water to drink, he steals Tsh. 2,000 and buys some candy to eat. He can't steal Tsh. 1,000,000—an amount such that my business dies." There's a loss any business has to accept.

The sale in wire and—when done with the proper discretion—the "discounting" of poles is an acceptable kind of loss for Tanesco. Such diversions support street electricians and Tanesco workers and ease residents' experience of network maintenance and extension. By contrast, the control of the meter should remain fully under Tanesco's institutional authority. In part, this is because of the advent of prepaid digital technology over the last decade, which has made meters increasingly expensive. To get a discount on a meter is to cut into Tanesco's already-weak profit margins.

But there is a deeper aspect to this prohibition. Expensive meters are valuable in and of themselves, but they also contain the value of future consumer use and payment. Tanesco cannot connect unregistered meters to consumer accounts, which widens the scope for tampering and hampers its ability to manage the system and plan for the future. This is why the meter is associated with the company's "inner secrets" or, perhaps better rendered, its "inner constancy" (Das 2014, 285; see introduction, note 16). As Dunky implies, it is a thing of the *umma*, the community, in the sense that it forms the elementary definition of the relationship between state utility and citizen. For Tanesco clerks and *vishoka*, funneling meters to residents for purchase is to cut into the system at its most critical inflection point and thus risk serious institutional sanction.

The meter is thus distinguished from wires or poles in that a customer must usually pay its full price and obtain it through the proper channels. But this threshold preserves the basic integrity of the service line extension while making it responsive to local particularities. It anchors the "real price," which is then tacitly adjusted through discounts and premiums on a variety of ancillary registers, including less-critical parts (poles, wiring), expediency, and bureaucratic compliance. This is not to say that meter theft does not occur, of course. There is reportedly a healthy trade in meters that have been stolen from neighborhood households or salvaged from demolition zones, which are then used as replacements, house extensions, or new connections. The point is that, whether inside the office or out, one has exceeded the baseline premise that defines the consumer-utility relationship.

And this brings us back to the case of the *kijana* and the Mama. The *kijana* was already somewhat beyond the pale by salvaging (stealing?) a meter. But as far as Besti was concerned, he might have at least mitigated the transgression by registering it under the Mama's name—a neglect his client ended up paying for. While we don't know to what extent the *kijana* himself "got away with it," the fact that he stopped picking up his phone indexes an ethos of short-term tactical improvisation that is not particularly sustainable in the long run. It suggests he is more force than form, not so much making a living as "living it up" from moment to moment. By contrast, Dunky suggests, sensible customers and/or *vishoka* should thus have the forbearance to restrict themselves to "small things." The desire for shortcuts is normal; the problem is indulging that desire to the point of self-negation, to the point that it forecloses its own conditions of possibility. By contrast, to reason modally is to think with and through the grammar of the power network, developing a feel for its force fields and pressure points, its trajectories, its "marginal gains" and losses (Guyer 2004).

cooperation and competition

As we have seen, the very term *kishoka* is contested; for some like Besti it categorically refers to the unethical. But for many others it simply refers to those "who go against official procedure" (*wanaoenda kinyume taratibu ya Tanesco*). *Vishoka* in this latter sense are a normal, even mundane feature of the network's infrastructural maintenance and extension. But while they operate within the network, they are never completely "of" it. In many ways this classically "parasitic" position is profitable, even productive: like many of the intercalary figures who have populated Africanist ethnography (Gluckman 1968; Kuper 1970; McGovern 2010, 184–90), they are mediators who, so long as they restrict themselves to relatively modest interventions, can make a living by bridging gaps within the Tanesco-consumer relation. One of the trickiest aspects of mediating service line extensions, however, is the complex mixture of collaboration and competition it engenders with Tanesco employees. While *vishoka* benefit from the gaps and delays in service, Tanesco employees sometimes push their own entrepreneurial schemes too far and infringe on the rights of the customer, not just delaying access to the service line extension or raising its transaction costs but obstructing it altogether. Such extortions may amount to simple greed. But I also think that, more subtly, it might reflect the way Tanesco employees may well come to resent *vishoka* as *arrivistes* who, while tolerable and even profitable to deal with, are ultimately beneath them.

We can explore the unstable mix of cooperation and competition in one final case study that concerns the legality of a wire, one whose complications unfold in such a way as to grant a sight line onto some of the social mechanics at work. As such, it will benefit from a style of analysis "that puts transactions into slow motion and lingers on every frame" (Guyer 2004, 18). The story takes place in an *uswahilini* neighborhood where the houses are built *kiholelaholela*, higgledy-piggledy. Besti narrates, and all the italicized words are English words and phrases.

> The house was there, I finished doing the wiring, and I sent in the application as is the standard procedure, because the poles were very close to the house [i.e., within 30 meters]. It came to the time that the surveyor arrived—and yes, there were problems as usual, so I had decided to pay some money to get the surveyor to come. The surveyor arrived, calculated the distance between the pole and the bracket where the wire was to connect. It was 23 meters; *I was quite within the range*, of its span, that wasn't a

problem. But next door there was a neighboring plot with a household. Then, while [the surveyor] was measuring the distance, the owner [of the neighboring plot] came over and said "I'm planning to build a [second] house here"—the very place where the surveyor was calculating the distance. At this point I told him—"that doesn't concern me, right now the space is open, and I'm certainly permitted to have electricity there." So, the surveyor finished his calculations and finished the procedure in a fine manner. I went and paid the Tsh. 55,000 [price for the surveyor].

However, at that point, unfortunately there was a shortage of meters. So, we waited like six months, and finally the meters came. I was on the waiting list which said I had already completed payments and was simply waiting for the meters to come. So, the meters came and one day I was called on the phone: "Are you so and so? Yes I am. We're the Service Line department and we will be coming to install your meter."

When they finally arrived, unfortunately, in the intervening period, the neighbor had already built his house. So when they arrived there they balked at installing the meter, insisting that the service line wire is prohibited from passing over another house. I insisted that that problem does not concern me, since I already paid and what I require is electricity. "The surveyor came and determined that I qualified for the service" [I told them]. But they balked completely. Now in their balking, there was a certain *aspect*: if I could provide some money, not less than Tsh. 200,000, they could install the meter for me—with full knowledge that they are disregarding the regulation. . . . I rejected that on behalf of my customer, telling him "no, don't relinquish even Tsh. 100." He was ready to provide the money, since he needed the service . . . but I said not even Tsh. 100, you have no reason to provide money when the office has already contracted to provide the service that you are acknowledged as needing. Consider how it would've been had the meter been installed before the neighboring house had been built! So this aspect of "legalizing" the installation of the meter through money, we won't be doing.

So, I rejected this and he didn't provide any money, so the team wrote a report that they didn't install and left. The next day I went to the office to *pledge complaint*. I said "this, this and that happened." I didn't discuss the money issue, I didn't discuss that they said "we need money," I didn't say anything because I didn't consider it important. All I simply said is that I want electricity. They explained to me I don't know how many times that "there are certain regulations" and I said "look I know the regulations, but the one who approved the electricity poles was whom? Your agent! He didn't know there was any housing structure there. And since I already paid *I'm entitled to service, of which I've paid for*. So gentlemen what I need is service."

So we went through these questions of this and that, and well anyway, they saw that I had solid grievance. They said, "look we haven't denied you, we will install the line and meter. Go to the neighbor and settle things with him, get him to write a letter that permits wires to pass over his property." And because I needed service, I didn't see any other way, so I went to speak with him. *Of course I had to seduce the guy*, to try and explain to him *that it's not risky*, you see? And that it even would be beneficial for him, since he himself hadn't connected his household . . . I had to *ask the chap that if he remains under that* [sic], there will come a day when Tanesco denies him service for causing them to lose a customer. *I had to play tricks of course, knowing that it wasn't right, but what had I to do? So by solving that*, he wrote a letter for me, he signed it, I sent the letter there, they filed it and included it in my customer's file. They came, and they connected his power. Now look, had that neighbor declined, [my] customer would incur extra costs, as Tanesco was insisting he purchase another electrical pole on the other side of the house . . . now, should it be the case that he didn't have the extra money for that, what would've happened? *Now, had I been unscrupulous as well I could've capitalized on the same problem*, and exploited a customer without good reason. And so the way I see it the basic problem of the whole thing stems from the company itself, Tanesco. They have a lot of broken policies. They like to play very much the *blame game*, but they themselves are the root cause. I'm speaking of course as someone who's been in the game for a very long time, dealing with many of these problems and fighting against them.

The virtue of Besti's story is to show the many complications in play when negotiating a service line installation. The nominal price for the meter installation is set by Tanesco, and had already been paid back at the office. But once at the house, the actors calibrate the "real price" across the scales of social status, time, and legality. But here things became rather tangled. From Besti's perspective, the service line's "real price" had already been paid. It included money to the surveyor for showing up quickly, and a six-month wait—five months longer than the maximum waiting period promised in the customer service charter. The Service Line department foreman disagreed and proposed another variable; Besti would need to "legalize" the wire running over the neighbor's house. But what exactly was the foreman offering, and what was the logic by which he offered it? This is the axis on which the social drama turns.

Because the service line crew's nominal paycheck comes from Tanesco, they cannot additionally charge the customer to install his or her meter—at least not directly. So they disaggregate the installation into a number of modifiable ancillary values, such as speed and discretion. They may charge extra for "gas" to arrive in a timely fashion, or for "drinking water" to work around an irregularity. Crucially, these criteria precede the actual labor, and so we may think of them as a kind of currency conversion (Maurer 2012). In paying for the crew to arrive at a job, or for them to agree to finish it, one is essentially paying extra for better access to the transaction—paying to pay. (Indeed, it's interesting to note that these ancillary payments are made in "liquids," since the *kishoka* partially liquidates the restricted asset of Tanesco service in order to make it more accessible.)

Unfortunately for the service line crew, their chances to cash out withered on the vine for six months. Even had he wanted to, Besti could not have paid the crew to come to the site. There was no meter available in the store—at least, we imagine, to someone of his unexceptional social status—and therefore no logical way to pay to expedite its installation. Besti was thus left to "pay" for the crew in the default currency—time waited. But time is an insufficiently liquid currency; it may be a standard of value, but it is not a means of value transfer. Besti could certainly not pay them *in* it. Instead, the crew took their time, in accordance with the default ordering of Tanesco's waiting list. In doing so, they forfeited their ability to charge for bumping Besti's customer up the queue. To arrive is, logically enough, to be too late to charge for gas money.

The action thus shifted to a second possibility. Once at the site, the foreman spotted an irregularity, and demanded "drinking water" to legalize it. But this too was too late. As Besti pointed out to them, the surveyor had in effect already legalized the line six months before, when there was no house built below it. On the one hand, Besti's bureaucratic logic might strike us as somewhat Kafkaesque. Despite what the surveyor's diagram said, there was a real, not to mention dangerous, wire over a real house. Nevertheless, the logic holds. The foreman could not legitimately be concerned about the actual danger posed by the wire, since this was the very thing he was offering to overlook. Thus: unable to charge for the performative speech-act of bureaucratic approval, since Besti already possessed it in the form of the surveyor's official diagram, and unable (or perhaps just unmotivated) to change the real-world conditions of the service line, what was the foreman charging for? We are forced to conclude: for nothing but a rent on the installation itself. The foreman's offer to legalize the line was in effect a demand for tribute, an exchange so tenuous that it could only be received as an insultingly naked and (in the negative sense of the word) parasitic fee on passage.

At this point however, a few questions come up. The foreman might have very well been successful in his ploy, coarse as it was. After all, the client was willing to pay the extra fee to "legalize" the wire. He failed to make any distinction between the various currencies of the transaction—between, say, the legitimacy of gas money versus a premium on the installation itself—he simply wanted to get connected and judged it worth the last-minute Tsh. 200,000. And Besti could have colluded with the service line crew and recommended this course of action. Indeed, he might have even received a cut of the Tsh. 200,000 had he done so. But he was dead set against it. Why should this be? It does not seem convincing that Besti firmly stands on categorical legal or moral principle exactly, since in other cases he was willing to pay the surveyor and "play tricks" on the neighbor. Supply had met demand, and it was by far the easiest option to let the customer pay, so why did Besti object?

At stake here, I suspect, is a question of personhood and the degree to which it is to be constituted by the search for short-term profit versus long-term social recognition. As his eloquence and sprinkling of educated English suggests, Besti felt himself to be a professional in an unprofessional milieu of "broken policies." The premium perhaps not only violated his customer's baseline right to access to the transaction, but also by extension, Besti's ultimate identity as a professional *contractor*, not just a fixer who is only looking out for the best score.

Obviously, this should not lead us to thinking that Besti is some sort of principled bourgeois gentleman, heedless of the importance of maintaining good working relationships with Tanesco. While Besti would certainly disdain to call himself a *kishoka*, which he idiomatically reserves for "quacks" and thieves, he is nevertheless part of the *kishoka* world as I have described it here. His recourse to the Tanesco office, which both refused and acceded to the problem of "legalizing" the wire, reflects as much.

In fact, Besti's actions conform to a logic that we have seen before. It is the logic of preserving the meter (and its installation) as an object whose price is fixed, thereby freeing its ancillary features to the play of the market. Besti was fine with paying the surveyor his drinking water. But he stood indignantly on principle when the foreman proposed to modify the nominal price itself. Just as Tanesco managers do not brook meter theft, Besti was not willing to pay more than its full amount (nominal cost + time waited), as this would have made him "unscrupulous" in relation to his customer—that is, in his view, a *kishoka*. Instead, he took up the conventional bureaucratic recourse of "pledging complaint" at the Tanesco office, even as he refrained from mentioning the failed extortions that necessitated it.

The broadest analytical point to come from all this is that the service line extension cannot be reduced to an official procedure nor to freewheeling market negotiation but rather exists as a dynamic interplay between the two. In fact, Besti's recourse to the Tanesco office bears striking resemblance to interpretations that Hart put forward in his Malinowski Lecture (1986, 649) regarding some details of Trobriand ethnography, specifically *wasi*, or ceremonial prestation controlled by Big Men, and *vava*, or individualized barter. The lecture deftly captures the "practical and conceptual complementarity" between gift and commodity exchange mechanisms often regarded as separate:

> A breakdown in political relations between coastal and inland villages (i.e. war) might occasion a shift from *vava* to the more formal *wasi*. Equally unpredictable fluctuations in supply (failure of the fish catch or a yam glut) might undermine the price-setting mechanisms of barter and require the intervention of big men and rationing or stockpiling agents. . . . Normal conditions grant low-level agents considerable autonomy which is superseded by high-level regulation when the environment is especially uncertain. Big men may affect to despise haggling—it is not, after all, their stock-in-trade—but we would be ill-advised to take at face value statements that formal exchange and ordinary barter have nothing to do with each other.

Besti's recourse to Tanesco is precisely a kind of supersession of "high-level regulation when the environment is especially uncertain." All the characteristics of formal exchange that Hart describes are in play: the ritualized bureaucratic procedures of complaint; the time-delay of waiting until the following day; the retreat from direct conflict with the service line crew, all of which attempt to phatically stabilize a case of market haggling gone awry. It is further noteworthy that, conversely, the Tanesco office "big men" themselves were less interested in upholding the regulation than minimizing their culpability. The team could run a wire over the house as long as the neighbor agreed to write a letter—an eminently pragmatic-political solution, and not a rational-legal one. At both high and low levels, then, negotiations took place within certain thresholds such that some viable midrange flexibility is sustained, and so that *vishoka* may be incorporated into the service line extension without it completely cannibalizing its basic logic.

conclusion: becoming infrastructure

Besti ends his story with an invocation of the long-running nature of his encounters with Tanesco, a weary but self-possessed professionalism. In this final section, I examine some of the ways in which contractor *vishoka* conceive of themselves as becoming infrastructure.

Here is Samueli, who earlier in the chapter expressed tentative hope for prospects "in the street" after losing his position at Tanesco.

> SAMUELI: I've lived this way for four years. The street pays more than the company. Since I've left I've been able to build two houses, whereas then even rent was a problem. . . . Since I've been fired from the company, I live better than the *vibarua* who are still inside. I live a good life . . . yes, the days are incomparable, there's Monday and then there's Tuesday, but on the days when it pleases God I can make Tsh. 200,000, even Tsh. 600,000. I can do wiring for up to 1,000,000. And with that, even if I spend the rest of the month just sitting and not working, I'm still doing better than when I was working inside the company. There the boss loads you with work, where sometimes you don't leave until 9 at night or 10, sometimes you're at the site until morning for Tsh. 5,000. If you protest to the boss "why are you having us work for more time than we're getting paid" he'll say don't come into work tomorrow. They just fire you. It's much better now than in the company.

> MD: If I was your fellow *kibarua*, and I'd seen that you'd gone into the street and were doing so well, I would ask myself: why don't I go into the street too?

> SAMUELI: You know with people, it's said that our stars are incomparable. Your star and mine are different. You won't have the capacity to communicate that I have. I can talk to the customer and tell him I will do his work quickly and be trustworthy about it. If you were to follow me, you might have a lust for money, or ruin yourself with drink or just not be capable of the work. You might not reach the level I have.

Samueli seems to have followed the star that is his particular skill set and achieved one of the highest goals in Dar es Salaam: to build his own house. But what sorts of skills are these? In her ethnography of high finance, Karen Ho (2009) identifies liquidity as the proper habitus of Wall Street traders: it confers upon them adaptability and flexibility, and means to profit from Wall Street's chronically short-term institutional and transactional horizons. It is interesting to note that people like Samueli, toiling a world away in Africa's chronically "unbankable" urban worlds, are often written about in the same way, as mercurial tricksters and hustlers, but also in effect as "infrastructure" (Simone 2004b), as "connected" and capacious nodes in a complex network. There is a great deal of truth to these depictions of fast capitalist *ujanja* in both the global center and periphery, and it is enlightening to consider their hidden affinities. Indeed, anyone who still imagines Africans to be trapped in a communalist, role-governed traditionalism will be surprised to hear Samueli's radically individualized vision of personhood, each its own star. *Vishoka* must be willing to aggressively pursue a living, often in competition with each other, in a milieu that includes various degrees of outright theft and con artistry.

But we can be more precise here. Wall Street traders may be liquid, but their class position is not, and this generally insulates them from career failure. By contrast, *vishoka* can only arbitrage their position to engage in antisocial, "entrepreneurial capture" kinds of transactions for so long. Their luck might hold, as it seemed to do with the *kijana*, but then again it may not. Because "the days are incomparable," a strategy oriented to easy money is inherently feast or famine. To smooth out that unpredictability, more ambitious *vishoka* like Samueli and Besti must resist giving free reign to that very ambition in their day-to-day operations. Against a habitus of liquidity, they forge a kind of ethical stability that is built around reputation, reliability, and clear communication. They seek to decommoditize parts of themselves and become solid: to escape the dispensability of commoditized labor, to caulk the financial dissipation of rent, to master the churn of lust, greed, or alcohol. In short they seek to become a kind of infrastructure, social agents that can control and direct such flows, rather than remaining exposed and vulnerable to them.

In interviews, *vishoka* often spoke in infrastructural imagery, as the "bridge," or "link" between customers and Tanesco, reliable to both. One former Tanesco *kibarua* named Abdallah observed:

> It depends how you were on the job. If when you were working there you were a bad person, he [inside] won't be helping you. But if you got on well with him and you arrive, you can say "hey *mzee*, brother, come here for a minute" and he comes . . . [but] you can't get to he who is inside. Even if you want to give him money he doesn't want it. He thinks you might be a policeman. But he knows me, a worker who he trusts . . . I give him the work and I bargain . . . then I tell the customer how much I'm going to charge. So he eats and you eat. He puts the materials, he gets all the people to do the work. And I'm like the bridge who gives him the money.

Yet as bridges, *vishoka* suffer from structural tension. On the one hand, the process of building up a personal reputation has the power to traverse distinctions between employed and unemployed. Abdallah is still a "worker" to his collaborators at the utility, precisely because he was a good person to them as a *kibarua*. However, the upward mobility afforded to *vishoka* can also breed resentment among Tanesco workers. Here is the self-described "contractor" named Simon who operated at the *kijiwe* (hangout spot; see introduction) opposite a Tanesco branch office, stamping service line applications and expediting them to employees on the "inside," across the street.

> It's true, [Tanesco workers] don't want to respect us. Although we provide a lot of benefit they don't—except insofar as it benefits them—respect us. There is because the way we Tanzanians despise each other, the way we don't value each other. In every part of life. Each person thinks only he should eat. Because this life that we have here [at the *kijiwe*], it can provide for us. I've been able to raise two children through this work. I'm building a house and I'm in the very final stages of it. A fine house! Well . . . it's not bad. It has three rooms, a dining room, a sitting room, master toilet, a public toilet—with just this work! My child attends primary school, although it's a government school. But my youngest child attends a Catholic nursery school, Tsh. 300,000 per year. This is expensive for us at the *kijiwe*! It's hard in our environment to make profit.
>
> But you find, still, a person can despise you. Even worse is if you decide to pass each other in a bar. They enter and we enter. They order meat and we order meat. They order beer and we order beer. And that bothers them. "Ah, those guys just sit outside, no official work but their cheeks are filling out, their necks are filling out, what are they eating?" And they regard us as what? "Those *vishoka*, those thieves, those saboteurs!" They judge us on appearances. If you have something to lick [*ukilamba*]: "where'd you get the money?" If you fill out, if you're fresh: "where'd he get that money?" They despise us. Based on how we look [sitting at the *kijiwe*].
>
> And there are people there that live fine lives. I've been there a long time and I've been late in building my house, but there's lots of us at the *kijiwe* that have their own places. Lots. They live normal lives; they school their children. Just for the work of sitting there at the *kijiwe*. But to plug into NSSF [National Social Security Fund] is difficult, to get [an] AKIBA [bank loan] is hard, since those are intended for salaried employees. But I can say it's still preferable. The way I see it, employment is slavery. You're under the law, under pressure and then caught up in scandals, fights, witchcraft. . . .

To be self-employed is better, and I pray to God he continues to grant me the ability to be self-employed rather than employed.

Substantively, successful *vishoka* are members of a professional community. They go to the same kinds of bars, build the same kinds of houses and send their children to the same kinds of schools as Tanesco workers. Yet there comes a point when their success throws their informal presence ("how we look") into relief.

In the predicaments of *vishoka*, there are remarkable echoes of socioeconomic relations as far back as the latter half of the nineteenth century. Low-status newcomers and dependents on the Swahili coast struggled to become *waungwana*, urban gentleman, and carve out spheres of economic autonomy for themselves, often in the face of reluctant disdain from patrons who nevertheless tried to profit from their strivings. Indeed, for Simon and Issa and surely for Samueli, official employment is one more form of *utumwa*. Their attempts to become their own bosses, in control of their own time, are tolerated by their patrons, but disdain for "*vishoka*, thieves, and saboteurs," peeks out from the veneer. Patrons in the precolonial era likewise regarded slaves as inarticulate beasts who could at best mime true membership in a proper social community (Glassman 1991). But slaves of course could communicate, and with varying degrees of success they incorporated themselves into their host communities.

In one sense, the "becoming infrastructure" of *vishoka* expresses a historical logic of personhood that has endured through Tanzania's state-centric socialist developmentalism and its neoliberal reaction. Since the 2000s, the receding tide of neoliberal triumphalism has revealed that markets, like theft, do not always need to negate public goods so much as modulate them. Tanesco remains a public monopoly with a developmentalist commitment to connect all residents. Beyond this baseline, however, market mechanisms in the form of *vishoka* have emerged to modify the speed, ease, reliability, and expense of connection. They tacitly transform the ways Tanesco consumers access electricity, both facilitating that access and constituting a tax on it. An ambiguous mix of bridge and toll collector, host and guest, *vishoka* illustrate how infrastructures become markets for private gain, while market transactions become important channels to access public goods and services. This ambiguous blending makes them inherently unstable figures, ones who are "in" but not entirely "of" the infrastructure of network extension. In this way they stand for the principle of modification itself, for its logic of both transforming and preserving, cutting and connecting. I have tried to render this logic ethnographically, the ways it unfolds through the bloom and buzz of deals and scams, the complexities of careers and personalities, and the historical experiences of dependence, autonomy, and obligation. In sum, I have tried to capture the sheer energy—social and technical—that weaves through the municipal electricity grid. The concluding chapter reflects on these energies in relation to the two central thematics of this book: postsocialism and infrastructure.

"Entropy is peaceful and democratic in the way that death is peaceful and democratic. Entropy means the ultimate death of the universe, or, in microcosm, of any closed system." I was attracted and repelled by that phrase: a closed system. Death, I thought, death in life; and I thought of growing older, growing old, in Dunn Street. The idea did not occur to me, even then, that marriage might yet be another closed system.   —VICTORIA GLENDINNING, *ELECTRICITY*

Two entities, call them "big L" and "little l," had to interact, such that each could leave an impression on the other. . . . At some point, they go their separate ways in a very peculiar way. Little l says, bind me and blind me, for I want to live. And big L says unbind me and unblind me, for I would rather die. You want big L, you want to live it up, you've got to give up little l. Conversely, you want little l, you want to live a long time, you've got to give up big L.
—PAUL KOCKELMAN, *THE ART OF INTERPRETATION IN THE AGE OF COMPUTATION*

conclusion

THE INGENUITY OF INFRASTRUCTURE

Victoria Glendinning's novel *Electricity* concerns the Victorian-era life of a petty-bourgeois woman named Charlotte Mortimer.[1] Charlotte lives in London during the 1880s, the same decade that the first electrified building in East Africa, the Omani Sultan's "House of Wonders," would light up the night sky—a dawning "Age of Electricity" that would connect metropole to colony in deepening relations of extraction (Rodney 1972, 179). In this epigraph from the beginning of the novel, the unmarried Charlotte finds herself stifled by the patriarchal dreariness of her natal household on Dunn Street. She is just getting to know an eccentric young man who studies the nascent science of electrical engineering. Charlotte senses that he and his strange new ideas might provide an escape to a new and fuller life—yet unaware that their marriage will turn out to be another stifling social system closing in on her. As Kockelman might put it, she is unaware that "to live"—to endure—always requires some measure of "binding and blinding." It is only later, after she has an affair, is widowed, and passes through a kind of social death does she truly "live it up," finding an evanescent second life as a spirit medium, performing seances in the London underworld with the help of some flickering electrical circuits. She turns, in effect, to a kind of *kishoka*-esque work, with all of its overtones of entrepreneurial capture and potential fraudulence.

Like her more famous literary cousin Dr. Frankenstein, Charlotte's encounter with the invisible and world-historical force of electricity dramatizes the twin liminal states of "death in life" and "life in death" that can beset any process of social reproduction. Over the course of her life she is alternately a body without an animating current, and a spirit whose surges exceed the social forms meant to bind it. Rarely does she find the right ratio. Such shortcomings and excesses always disrupt the dream of a perfectly functioning technical system, or a perfectly functional social structure. Against the inevitability of social and natural entropy, Charlotte's biography thus poses the question of the good-enough life. And as Kockelman's parable intimates, this is a question of economy. What does one have to give up in order to "live it up," and what does one have to give up in order to just live?

Current and wire, desire and discipline: this is the coupling of vital energy and social form that also animates the complex ethics of "conviviality" in African worlds (Nyamnjoh 2017). Though it unfolds with much different gendered and historical dynamics, Tanzanians too must find ways to align the forces of collective existence—electricity, money, desire—with the social forms that channel it. As chapters 3 and 4 explored, some residents during my fieldwork "lived it up" by pirating current, but at the cost of breaking the very premises of social interdependence that made that current possible. Others resigned themselves to "playing by the rules" and enduring basic services that were expensive and unreliable. Many residents tacked to somewhere in the middle. They might allow themselves to passively run up debt on the meter or surreptitiously reconnect after a Tanesco DC team rolled through and cut them off. Such hedged bets mark the point at which state-sponsored visions of development found provisional traction with residents' evolving practices of urban life, from Swahili-style houses with their buzz of tenants and kin networks, to the enclaves of the city's new upper middle class. Sometimes they gave rise to poignant social dramas of recrimination: who was not upholding their commitments to "society," to Tanzania's shared national project? Most of the time, however, these compromises remained hovering below the surface of political contestation—allowing the network to adapt and evolve along with the city itself.

These compromises were themselves responses to shifts occurring upstream, at the elite level of Tanzanian business and politics. Chapters 1 and 2 explored how, faced with the pressures and possibilities of liberalization, state and party elites leaned on sources of financial, political, and indeed literal "emergency power." Here too was a series of modal variations: continuous cycles of blackouts hedged by rationing, graft hedged by bailouts, price raises hedged by corruption sweeps. Together they held rule and legitimacy, current and currency, in contrapuntal tension, and set the scene for the street-level modifications below. In this way, electricity became unexpectedly central to a kind of second chapter in Tanzania's postsocialist transition in the 2000s, an interregnum with new kinds of continuities and breaks with the socialist past.

what was neoliberalism, and what comes next?

Much postsocialist scholarship has observed that beneath the formal ideological ruptures and embrace of liberalization, more substantive structures of socialist life endure (Askew 2006; Pitcher and Askew 2006; Schwenkel and Leshkowich 2012; Collier 2011; Harms 2011). Elizabeth Dunn's ethnography of the Polish meatpacking industry, for example, examines the way its incorporation into the technological zone of European Union–wide industrial standards ironically re-creates the very black markets that developed under the strictures of socialist discipline. Today, she observes, we see that the same kinds of persons flourish beyond and beneath EU regulatory strictures as under socialism: "Socialism's persons were 'hunters' who used their networks to evade the blockages created by the planned economy. The ability to dodge regulations, to get goods that the state denied, and to covertly sell the fruits of one's private labors became highly valorized as *spryta*, or cleverness. People spoke admiringly of exploits in evasion, making a snaking gesture with their hands to demonstrate the agility with which rules were circumvented" (Dunn 2005, 188).

This description, which could be taken almost word for word and applied to the residents of Dar es Salaam, down to the same terminology of agile cleverness, or "ingenuity," begins to pave the way for a wider comparative ambit of global postsocialisms (Rogers 2010), as well as an invitation to think it in relation to that other post: postcolonialism (Chari and Verdery 2009). Like Polish peasants, or the "urban hunters" of late 1990s Ulaanbaatar (Hojer and Pedersen 2019), residents of Dar es Salaam valorized an ethic of *ujanja* originally cultivated under the difficult shortage years of the 1970s and 1980s, if not earlier, under the colonial regime.[2] This ethic has served them well in forming parallel markets that supplement Tanesco's austerity programs and disciplinary reforms. We might even say that socialism has taken revenge for its supposed world historical defeat. Its model of entrepreneurial personhood is at the very heart of the post–Cold War dispensation of transnational capital. Hunters of all kinds comprise a shadowy "stealth of nations" (Neuwirth 2011), trekking around the world and propagating forms of "low end globalization" (Mathews and Yang 2012) that will be familiar to anyone who has wandered through electronics markets in Lagos, Guanzhou, Hong Kong, or Conakry, to say nothing of the far murkier trade in arms, organs, or people (Nordstrom 2004). These networks should give pause to the technocratic, air-conditioned confidence of European Union regulators and Western development donors alike.

The ideological fallout is even more pronounced. Whether due to the expansion of war and terrorism after September 11, 2001, or the financial crisis of 2007–2008, the hubris of an "End of History," free-market capitalism hardly needed the anthropology of postsocialism to debunk it. The pink tide of Latin America, the industrial hegemony of China, the anti-IMF sentiment of India, and the "late socialism" of Vietnam all outflanked the standard version of liberal-democratic capitalism. These trends made their way to Tanzania as well. By 2005 Tanesco's privatization had been rejected, CCM played to resurgent nationalism, and anti-Western sentiment prevailed. And yet I have tried to suggest that this too was something of an ideological feint, since the West was hardly synonymous with capitalism as it had been during the Cold War. Just as socialism was never the heroic string of Five Year Plans it purported to be,[3] neither has *uwazi* (liberalization) and *utendawazi* (globalization) quite lived up to their advertising. The reactionary invocation of an older socialist/nationalist posture by CCM can be seen as a symptom of the deeply but unexpectedly capitalist changes that *have* in fact come to Tanzania. Paraphrasing Brecht, we could ask: what is the rejection of Tanesco's privatization compared to its surreptitious, *de facto* privatization for the benefit of political oligarchs? Kelsall and Cooksey (2011, 35) well capture the spry agility of any ambitious political entrepreneur in Tanzania:

> During the process of creating and strengthening patronage networks, participants find out who they can trust. Individuals with the most ambition and determination become patrons, while those who are either too timid or (especially) principled get left behind. An aspirant *rentier* must be prepared to spend many years lobbying and ingratiating him/herself with existing patrons in order to rise up the slippery pole of clientelism. This process of network creation ensures that the most ruthless and single-minded actors come to dominate rent-seeking processes of all kinds.

And yet precisely because the qualities of ingenuity, agility, and cleverness are bound up in forms of accumulation, they cannot evade popular critique. As Janet Roitman (2005) illustrated for the Chad Basin, new kinds of wealth and work are inevitably accompanied by debates concerning their moral and political regulation. When *vishoka* are officially denounced as thieves or saboteurs, electricians crack wise that *vishoka* (whoever they are) are nothing compared to the *vishoka papa*, literally the shark *vishoka* with access to government contracts and influence. Or to use another animal metaphor, they observe that *mtoto wa nyoka ni nyoka*—the child of a snake is a snake. In other words, they frame themselves as logical derivations of the "bad" Fathers that have gorged on the national cake, and suggest that their behavior is more than justifiable in light of the hardships the leaders have caused.[4] David Harvey's (2007) pithy history of neoliberalism is one in which the ruling class managed to liberate itself from the demands of the working class by capturing the state. Here we could simply reshuffle that idea a little; *vishoka* morally "capture"—or at least mirror—the ruling political class by "liberating" themselves from odious state regulation.

At the same time, beneath these recriminations there is an even more interesting pragmatics at play that acknowledges some dynamic limit to capture and redistribution. As the Richmond and IPTL scandals showed, the proverbial 10 percent on procurement contracts is one thing, but accumulation beyond it might be more than the body politic can handle. At the street level too, residents and electricians drew distinctions between diverting current or currency in ways that are relatively socially productive versus those that are not. Historically, the polarized alternatives to such compromises have been the miseries of the state's command economy in the 1980s and the savageries of a winner-take-all capitalism that has constantly threatened to succeed it. As I suggested in chapter 2, this is the unappetizing choice between collective hunger and maldistributed satiety. Still, by the end of Kikwete's second term, life in Dar es Salaam had in general shifted toward the latter, for better and worse.

what would magufuli do?

When I returned to the *kijiwe* in the summer of 2016, I spent a long and happy afternoon catching up with Simon, Dunky, and many others who were still around, still plying their mixture of contracting, stamping, and the occasional *kishoka*-esque work. Eventually, the evening rolled around and everyone made their way to the *daladala* stops to begin their series of punishing bus rides to their homes and families in neighborhoods like Tungi, Buguruni, and Gongo L'Mboto. Soon it was just myself and another friend, a charismatic *bodaboda* (motorcycle taxi) driver who was, I always enjoyed noting, named after an equally charismatic socialist leader of the twentieth century—I will call him Castro. Castro also counted the *kijiwe* as a place of employment, giving rides to Tanesco and other government office workers in the area. But, as his own late-evening lingering suggested, business had been bad since late 2015, when authoritarian, anticorruption populist John Pombe Magufuli of CCM won the national election. "Under Kikwete people had the means to make money," he grumbled. "But JPM is making it hard for people who do their work *ki-mission town* (mission town style), and for *wale wa chini* (those at the bottom) who rely on *wale wakubwa* (the big ones)." Whereas relatively comfortable bureaucrats working at the government ministry could once pay him Tsh. 20,000 for a ride to the Ubungo bus terminal in the north of the city, JPM's crackdown on graft meant that they were now being "forced to live within their salaries" and could only pay Tsh. 5,000. Even Dar's interminable traffic had eased, Castro observed, since people had no money for fuel. "Before, a girl/mistress could call up her man and say 'baby, there's no gas for the car,' and he'd say 'ok take Tsh. 100,000 or 200,000.' Now the car stays at home."

As discussed in chapter 2, the 2015 presidential race in Tanzania reflected the tensions of an incumbent ruling party dealing with decades of liberalization and the associated sapping of its moral legitimacy—all against the background of yet another round of power rationing. It ultimately produced an outcome that intensified some dynamics of the Kikwete years and dramatically countered others. When the votes came in on October 25, the ruling party had once again prevailed, though allegations of vote tampering did not go unvoiced.[5] With surprising zeal, Magufuli implemented his promised campaign of moral rectitude. In December he cancelled independence day celebrations and directed the money to hospitals fighting a local cholera outbreak. He subsequently banned government spending on holidays and celebrations, as well as the lucrative "per diem" economy of workshops and trips abroad for state officials, and eliminated fake "ghost workers" on government payrolls. Tax collection increased dramatically, and water and electricity debtors, including government institutions and Zanzibar, were threatened with both disconnection and Magufuli's "wrath."[6] Eventually, even the owners (and inheritors) of the controversial IPTL contract, widely assumed to be protected under Kikwete's presidency, were arrested on the classically socialist charge of "economic sabotage."[7]

In the early months of the Magufuli presidency, this "lancing of boils" was incredibly popular. Even Ally, caught up in the early optimism, reflected that "only four [African] presidents have truly understood the *uchungu* [bitterness; i.e., the suffering] of people: Lumumba, Mandela, Sankara and Magufuli."[8] A popular twitter hashtag asked #WhatWouldMagufuliDo?, gesturing to the way the new president had become an exemplar of moral leadership. Kenyans in particular drew comparisons between their own president's lavish ceremonies and accoutrements and Magufuli's spartan style. At the same time, however, there was indeed something "bulldozing" about these policies, and the hashtag soon lurched into a kind of absurdist humor. One tweet posted a picture of a lit candle suspended beneath a shower head, with the caption "Needed to take a hot shower but electricity too expensive, then I thought what would Magufuli do?" Other cost-cutting improvisations included a hairbrush replacing a car's broken side-mirror, or loaves of bread in place of a wedding cake (Brankamp 2015).

At the center of these ambivalent and ambivalently self-mocking sorts of praises was an uneasy intimation of the right measure. Magufuli was reprising a kind of Nyererean austerity, wherein discipline and punishment would be redeemed by a proper developmentalism. But after decades of deferral, was this redemption plausible? It was one thing to be clever and cost-cutting—a certain level of it aligned with the socialist virtues of self-reliance (*kujitegemea*). And yet as many Tanzanians remembered, self-reliance could be a euphemism for failed development, even de-modernization. Indeed, after a certain point austerity did not just breed ingenuity but hunger—this was Castro's point. Whatever else it did, the rent-seeking of the Kikwete years had created an ecology whereby the "small fry" were able to live off the largesse of the "big shots": a whole redistributive economy of debt and deferral had flowed down from institutions like Tanesco through its employees and to their unofficial auxiliaries—including the *vishoka* and *bodaboda* drivers para-sited at the *kijiwe* across the street. Now that flow was contracting. "Those who support him do their work and shut up," Castro shrugged. "For those that don't, it's *ki-mission town*!"—that is, a mode of clever hustling whose returns were getting harder and thinner.[9]

To some observers, not least the educated urban critics that formed the core of popular opposition to CCM, the state's disciplinary posture was not indicative of a renewed moralism but rather of a consolidating dictatorship (*udikteta*). Some policies of the new administration harkened back to the state's most intrusive attempts at regulating gender and sexuality (Ivaska

2011). In 2017, Magufuli endorsed a 1960s era law that licensed mandatory pregnancy tests for schoolgirls and expulsion for those who "failed" them.[10] The same year, Dar es Salaam regional commissioner Paul Makonda threatened to round up and arrest LGBTQ+ residents—possibly to strengthen Magufuli's bona fides with the country's growing evangelical population.[11] Political repression deepened as well. Alongside the Cybercrimes and Statistics Acts that had been passed in early 2015, the state extended a ban on opposition rallies and live broadcast of parliamentary debates. The 2012 kidnapping and beating of striking doctor Steven Ulimboka mentioned in chapter 2 foreshadowed an expanded campaign of violence, with dozens of opposition activists disappeared or murdered by *watu wasiojulikana*—unknown people.[12] Tundu Lissu, the charismatic CHADEMA leader, was shot sixteen times in a failed assassination attempt and fled the country.[13] More so than the flirtations with political crackdown that characterized Kikwete's presidency, the subsequent Magufuli era exemplified the latent vulnerability of a legacy ruling party trying to retain power in a multiparty, liberalized system. In the runup to the 2020 election, the violence only increased, and CCM's transformation into a fully authoritarian regime seemed decisive (Paget 2021).

But the Magufuli era proper ended, unexpectedly, amidst the swell of the COVID-19 pandemic. Along with fellow authoritarian leaders like Trump and Bolsanaro, Magufuli had almost immediately launched into various sorts of COVID denialism, asserting that "Coronavirus, which is a devil, cannot survive in the body of Christ.... It will burn instantly."[14] He also cast doubt on the efficacy of laboratory test kits, alleging—with a certain absurdist panache, it has to be said—that he had secretly sent in samples taken from a pawpaw and a goat that came back positive. By May 2020, the government had stopped publishing statistics on infections or mortality rates. Over the next few months he declared the country coronavirus free, mocked neighboring countries' mask and lockdown policies as capitulation to Western influence, and refused to consider any vaccination program. At the same time, the government also promoted herbal medicine and steam inhalation treatments for those with breathing difficulties, even establishing some at Muhumbili National Hospital.[15] Friends like Ally and Thierry spoke to me carefully over the phone about these mixed messages, averring that they personally had not seen or known anyone with the virus, even as grainy videos of night burials and crowded hospitals circulated throughout Tanzanian Whatsapp networks. Public-facing Tanzanian social media often took a page from the president's rhetoric and decried, not inaccurately, the hypocrisy of Western imperialists criticizing an African country when their own were failing so badly. Here then was a strange and distorted echo of self-reliance, of a society rolling up its windows to the outside world.

Then, in late February 2021, the president suddenly stopped appearing in public. His death was announced a few weeks later due to heart complications, and Vice President Samia Suluhu Hassan assumed office thereafter. It is unclear, as of this writing, what direction she and the party will take. While there is some indication of falling in line with international norms, at least in regards to the pandemic,[16] political repression seems to be continuing apace.[17] As might be expected, all manner of rumors and speculation swirled about Magufuli's disappearance and death, including the possibility that he had been poisoned by a jealous and resentful Kikwete. Opposition voices claimed to have evidence that it was a coronavirus infection, which would have a certain poetic justice.

the julius nyerere hydropower station

And what of electricity? Is the "emergency power" era over as well? Though no EPPs are currently contracted to the grid, total supply still remains just barely above peak demand, and periodic load shedding is still common. Since 2013, numerous IPP projects both renewable and otherwise—wind in Singida, coal in Rukwa—have phased in and out of various stages of implementation, a testament to the still-decentralized and uncoordinated investment environment. In late 2015, Tanesco did complete construction on the first of the Mtwara pipeline-fed, 150 MW natural gas power plants Kinyerezi I, followed in 2018 by Kinyerezi II, a 240 MW expansion, which has partially improved the consistency of the power supply.

Consumer renewables are becoming more prevalent as well. When I first spent a summer with Ally and his family in their unelectrified house in the peri-urban outskirts, we powered our approximately four-hour nightly use of lights, radio, and television by two car batteries connected to an inverter (see figure C.1). Each morning, I would help his teenage daughters pile the heavy batteries into a pull cart and walk them to a garage on the main road, which benefited from a service line connection to the national grid. Like other households in the neighborhood, we would drop the batteries off for the day, paying around Tsh. 2,000 to charge both (a cost my own room and board fee helped defray), and bring them home in the evening, relaxing to nightly installments of a Filipino telenovela over dinner. When I spent another summer with Ally in 2016, the house now featured a small "pico solar" system, useful for a few lights and phone charges—but no TV (see figure C.2).

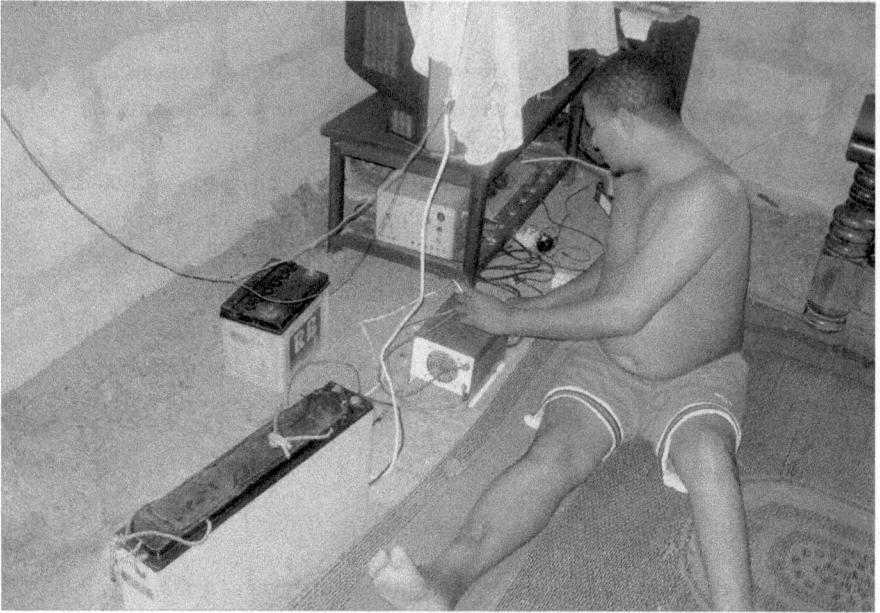

FIGURE C.1 *(above)*
Ally demonstrates his car battery-powered electricity setup, 2010.

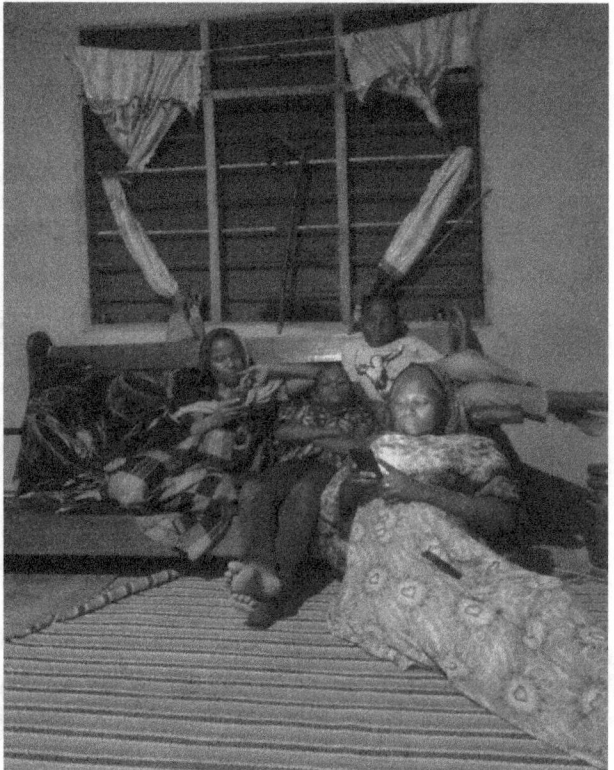

FIGURE C.2 *(beside)*
Ally's wife Amina and children relaxing at night, 2016, their solar battery system only strong enough for lights and phones.

Ally's family's dimly lit living room was a testament to the increasing affordability and commercialization of solar. Over the past ten years, photovoltaic panel technology has markedly improved, and Ally and other consumers in urban centers like Dar es Salaam have access to a range of household battery and panel systems. East Africa is host to a number of ambitious Silicon Valley startups, often partnering with NGOs or plugging into "development finance" to sell standalone setups to far flung rural villages on credit and with "pay as you go" technology (Rolfs et al. 2015). As I have described elsewhere, such systems offer the promise of energy autonomy and leapfrogging an older, centralized model of collective provision (Degani, Chalfin, and Cross 2020). A number of issues bedevil this promise, however, including systems of unpredictable or fraudulent—*feki*—quality, and their still limited ability to scale (Dean 2020; Phillips 2020). Across the continent, rural customers must wait for technicians to repair easily broken equipment, while adhering to repayment schedules often remains prohibitive (Adwek et al. 2020), sometimes resulting in fraught confrontations between households and company agents reminiscent of those I witnessed on DC patrols. Urban households, meanwhile, often use solar power in tandem with state connections. Contractors like Simon now regularly install automatic backup systems in wealthier homes for a seamless transition when grid power fails, but most households must mix and match in more uneven ways. When I returned once again to visit Ally and his family in 2018, they had at long last wrangled a Tanesco service line, ensuring prodigious amounts of television broadcast into the living room daily—if the supply held. If not, another small solar system (the previous battery had rather disturbingly warped and swollen) stood by to return the household to a more limited mode of electrification.

Significantly, Ally's new Tanesco line was a kind of reconnection. In the early 2000s, when their eldest was still a young girl, Ally and his wife Amina had in fact enjoyed a grid connection while renting a room in a Swahili compound in Temeke and were part of the wave of cutoffs (and surreptitious reconnections) that accompanied the NetGroup contract. They moved out to a plot of land along the peri-urban edges of the city a few years later, affordable in part because of its lack of infrastructural provisioning. Their lives over the years were thus marked by a waxing and waning of electrical current, and a shifting, sometimes ingenious assemblage of elements to channel it. Their experience, common to residents who resettle along the ruralized edges of Dar es Salaam, suggests that Africa's energy future will not lock into any one technology or mode, but rather continue to involve a diverse portfolio of resources and arrangements designed to manage uncertainty.

At the same time, Ally's experience also points to something broader about the persistence of national grids as technologies of political community and sites of affective investment. In a candid (?) moment at an "African innovators summit" in 2015, Ory Okolloh, Kenyan tech investor and creator of the Ushahidi platform for tracking political violence, offered a blunt assessment of what she called the "fetishization around entrepreneurship" in Africa:

> Like, "don't worry that there's no power because hey, you're going to do solar and innovate around that. Your schools suck, but hey there's this new model of schooling. Your roads are terrible, but hey, Uber works in Nairobi and that's innovation!" . . . I think sometimes we are running away from dealing with the really hard things. And the same people who are pushing this entrepreneurship and innovation thing are coming from places where your roads work, your electricity works, your teachers are well paid. I didn't see anyone entrepreneur-ing around public schooling in the US. You all went to public schools, you know, and then made it to Harvard or whatever. You turned on your light and it came on. No one is trying to innovate around your electricity power company. So why are we being made to do that? Our systems need to work and we need to figure our shit out.[18]

Of course it is arguable that veranda-dwelling elites and air-conditioned experts in the United States and beyond are very much promoting dubious "innovation" around access to schools and energy.[19] Across diverse ethnographic contexts, anthropologists have begun looking afresh at seemingly boring things like paying taxes and interacting with civil servants as lively, contested sites where the social contract between citizen and state is materialized—and the ways that public-private partnerships, market logics, and economization refract but never fully efface their moral, even utopian charge (Bear and Mathur 2015; Venkatesan 2020). Often deeply entwined with fiscal policy and administrative process, energy infrastructures are likewise sites of "ethical constitution" (High and Smith 2019). Their flows *substantiate*—that is, function as both sign and substance of—the social contract, or offer evidence for its reevaluation. As Soumhya Venkatesan (2020, 143) writes of tax, they are "indicative of ideas about what 'society' is, what constitutes 'the public good,' and where and to whom one's responsibility lies."

In Tanzania, Magufuli's energy policy said much about his vision of political community. Its centerpiece turned out to be neither privately generated gas nor commercially sold renewables but hydropower, and in particular the massive Steigler's Gorge dam. The dam site is located on the Rufiji river in the Selous Game Reserve, a World Heritage site and home to a dense array of wildlife—including a majority of the continent's dwindling elephant population. The project was originally proposed in the colonial era, and Nyerere vigorously supported it during the first decades of independence. In 1975 he formed the Rufiji Basin Development Agency, explicitly modeled on the United States Tennessee Valley Authority, to develop the project (Dye and Hartmann 2017).[20] The project languished through the 1980s and 1990s after the World Bank refused to allocate funding due to environmental concerns and, consonant with the neoliberal turn of the period, deemphasized megaprojects. It was resurrected in Kikwete's first term as a way to solve the persistent power crisis. Bids from Chinese and Brazilian developers were briefly entertained, only once again to be abandoned after the turn to gas generation after 2010.

The dam has faced objections both domestically and abroad. Like other projects of its size and ilk, it would exhibit drastic impacts on regional ecologies, populations, and livelihoods (Isaacman 2005; Tischler 2014). Moreover, while the massive 2100 MW project would by itself constitute more than the power network's entire installed capacity and peak demand, its reliance on consistent rainfall means it would be vulnerable to the same climatic fluctuations as the Mtera, Kidatu, or Kihansi dams. It would hardly be the silver bullet to Tanzania's power problems, nor is it clear that it could even bring down the average cost per kilowatt hour.

"The Bulldozer" remained undeterred. Despite no external donor support, a tender for the project's construction was awarded to Arab Contractors Limited of Egypt in 2018. During the ceremonial laying of the foundation stone the following year, Magufuli gave a speech that drew on the nationalist depth and resonances of the project.[21] It was, he observed, a "singular and historic day," one that Nyerere had begun dreaming of over fifty years ago, and a declaration that "our country is free and has the capacity to manage its own affairs as it sees fit." He was proud to say that the dam was financed entirely by the republic, through the tax collected from citizens; that it would be built by Egyptians, fellow Africans whose nationalist leader Gamal Abdel Nasser enjoyed a close personal friendship with Nyerere, and whose own ancestors had built the pyramids; and that it would provide reliable and affordable electricity, power his administration's industrialization drive, increase foreign and domestic investment, reduce local consumption of wood and charcoal, and become a hub for the local economy and ecology ("animals will come to drink, people will come to farm and fish"). Yes, he acknowledged, some critics have opposed the project. There are those companies that own hunting blocks in the Selous Reserve; how many Tanzanians were even aware that tourists stayed in $300-a-night hotels out there, flying in and out on private planes? Did his cabinet even know how much tax was collected on that business? Did his security and intelligence services even know how many resources were being smuggled out?[22] This is why he was ordering his Minister of Tourism to allocate half of those blocks to the formation of a new Nyerere National Park, where it would no longer be legal to "kill our animals." And the dam? It was (of course) to be called the Julius Nyerere Hydropower Station.

The speech was not just an overt "return to Nyerere" (Fouéré 2014) but a swelling of Tanzania's socialist brand of high modernism: an insistence on self-reliance and Pan-Africanist solidarity in the face of lingering colonial presumptions to tutelage; on a redistributive politics in which the state acts as the central governing intelligence; on the expulsion of parasites feeding off the public good; and, if only tacitly, on the necessity (for some populations) to sacrifice for that good.

The wide-ranging speech also managed to capture something of the dynamic cascading ecology of the project, at least as Magufuli imagined it: rain moving from sky to river to reservoir; the subsequent growth of irrigated agriculture and commercial fishing; the fall in hunting and rise in nature tourism, and of course the flow of smooth, clean current to a rapidly industrializing country. In the next section, I'll suggest that one important task for the anthropology of infrastructure might be to gather the conceptual resources to think with such cascades, and with the actors who attempt to harness, obstruct, or otherwise distribute their forces.

on modes

As a novice ethnographer, I became interested in studying electricity because it turned out to be such a surprisingly obvious medium of collective life. Learning Kiswahili during my first summers in Dar es Salaam in 2007 and 2008, I couldn't understand much of what was being said, but I could understand when the lights went out. Most obviously, this was a lesson in the "fragility of things" (Connolly 2013)—the way electricity is only as consistent as the network of wires, meters, and transformers it flows through. But it was also easy to see how its absence filtered the mood of social life, the way my host family grumbled as they fetched kerosene lamps, and eventually settled into candlelit dinners, jokes, and conversations. I thought back to my childhood in South Florida after a few especially bad hurricanes knocked out the power in different neighborhoods: how quiet my room was without the humming of the AC; the way people drove around to different public spaces to find food, outlets, information, and gossip (cf. Rupp 2016). Cut the power and the whole bloom and buzz of social life turns in "shifting cascades of collaboration and complexity" (Tsing 2015, 157).

We might understand this dynamic quality in terms of emergence: a whole with properties irreducible to its parts. A rich literature on complexity theory and the new materialism has theorized emergence through notions such as the assemblage, the vital, and the semiotic (DeLanda 2016; Bennett 2005; Kohn 2013), paving the way for a posthumanist recuperation of the world beyond symbolic representation. A whirlpool emerges when flowing lines of current find each other in a way unique to their material properties (Kohn 2013, 168), as does a hurricane, as does the proverbial "whirl of the organism" (Cavell 1969, 52; see also Ingold 2015, 7). But in some ways, the circum-human quality of emergence harkens back to the foundations of anthropology. Durkheim inaugurated his sociological method by insisting that collective life had a quality over and above that of the component individual ([1914] 2005). When people come together in a ritual, a crowd, or a public, they create, by virtue of their collective accommodation, a kind of electromagnetic field (see Singh 2014, 174; Mazzarella 2017, 52). Ingold (2015, 6) suggests that Henri Matisse's painting *The Dance* exemplifies this collective emergence, a roundel in which individual bodies are gracefully joined together in a self-spinning circuit. In emergence, what matters is not the simple aggregation of people and things, but whether, through their interactions, they *form* something. As Deleuze asks: if you plug it in, does it work . . . does something come through? (quoted in Boyer 2015, 531).

Remaining within the grace of emergent form is thus not so much a foregone conclusion as it is an achievement. During fieldwork, I'd often visit Ally's taxi stand, a shaded downtown spot beneath a large power transformer, where he would wait for customers, read the newspaper, and shoot the breeze with his friends and colleagues. Like the vibrant little *kijiwe* across from the Tanesco office, here too was a stupendous variety of ambulatory vendors, bearing everything from snacks to shoes to rat poison for sale. But Ally always kept his eye out for *wagawaha*, coffee vendors toting a coal-warmed kettle of dark strong coffee in one hand and a watery bucket of ceramic demitasse cups in the other. Whenever I stopped by, Ally would flag one down and treat me and his friends to a few cups.

I came to deeply appreciate this ritual, both as a gesture of Ally's friendship and for the way it eased me into the proverbial *baraza*, the "table" around which men (and the occasional woman or two) gathered to discuss the issues of the day. Each cup was blessedly cheap, just a few cents, and Ally relished playing the patron, fluidly gesturing for another pour when all had completed the round. In the beginning, however, I was unused to the scalding hot temperature of the demitasse servings, and waited for mine to cool. This inevitably put me about half a cup to a cup behind, and my dawdling rippled through the whole ecology of the scene. Ally, ever a gracious host, would politely wait for me to finish, once or twice asking a customer to wait so that he could settle the bill. The *mgawaha*, who moved about the city in his own circuits and with his own regulars, would sometimes glance at me impatiently as well. Lest I suck all the pleasure and elegance out of the ritual, I'd set about to "working" the cup. I'd discreetly swish the liquid, softly blow across the surface, nurse small sips and finally, as the temperature cooled to a point I could stand, down the rest—only to be served up the next one. While these never allowed me to catch up, they did minimize the lag to what I hope was a tolerable proportion.

Kockelman (2017, 109) playfully describes the thermos as "the exemplary medium" since it preserves "differences across distances by insuring that its contents stave off thermal equilibrium for short periods of time." But thermal distinction can work a little too well. It suited me to sand it down, opening up the cup to circuits and processes beyond itself—namely my breath and the wider air. For his part, Ally had to improvise the social performance, tolerating the fumbles that may have adulterated the pleasures of playing host. But these variations are what contained the ritual and ultimately allowed us to go on doing it. Over time we became practiced companions, our drinking habits aligned. I was reminded of a *methali* (proverb) Nyerere (1968, 5) invoked to demonstrate the inherently socialist nature of African life, but that seemed to apply to sociality as such: *mgeni siku mbili; siku ya tatu mpe jembe.* Treat your guest as a guest for two days, on the third day give him a hoe. One way or another, it takes work to be a good guest—to forestall its degradation (Kockelman 2016, 355–56).

Anthropologists often begin fieldwork as parasites. They are para-sited, positioned alongside the flow of life as both guest and nuisance, social stranger and social child. Ideally they incorporate themselves into that flow over time, and reciprocally alter it in ways that are just and generous. This methodological commitment of time, of the willingness to learn and be transformed (and to look foolish while doing it) also implies a corresponding intellectual attitude. "The people who make our research possible" Faye Harrison observes with powerful clarity, "are much like us. They have knowledge, sometimes very sophisticated understandings of the world, and embedded in those understandings are what we call theory" (Harrison et al. 2016).

In this spirit of "convivial scholarship" (Nyamnjoh 2017, 267), a key strategy of this book has been to tack back and forth between two theories of the parasite: one rooted in Michel Serres's work (and to some extent Kockelman's inventive appropriation of it via American strains of pragmatism and semiotics), whereby the parasite is an interruption of a relation—and a Tanzanian one that asks, essentially: who eats, and at whose expense? Both theories are hybrids of high and low. Serres draws on a European "politics of the belly" (Bayart 1993)—folk tales about guests, hosts, and meals—to theorize systems and relations, just as Nyerere interweaves energo-metabolic Kiswahili proverbs with Fabian socialist theories of exploitation and development he absorbed while studying at the University of Edinburgh in the early 1950s (Molony 2014). More broadly, I have explored how what we should indeed take seriously as social theory in Tanzania bears elective (electric?) affinities with a range of anthropological traditions broadly concerned with vital force and social form: the unresolvable interplay of rights (*haki*) and peace (*amani*) (chapter 2), desire (*tamaa*) and restraint (*utaratibu*) (chapter 3), or strangers/ dependents (*wageni*) and the community (*umma*) (chapter 4).

Such an analysis, I hope, results in a modestly decolonial sensibility in which my interlocutors are neither explained away nor reified as emissaries of unbridgeable, ontological difference. As Achille Mbembe notes, any properly decolonizing project must begin with the critique of Eurocentric knowledge production and a concomitant embrace of multiplicity (2021, 79). At the same time, a danger lies in reducing multiplicity to difference, in the fantasy of putting everything (back) in its proper place. "The concept of Africa invoked in most discourses on 'decolonization,'" Mbembe tartly observes, "is deployed as if there were unanimity within Africa itself about what is 'African' and what is not. Most of the time, the 'African' is equated with the 'indigenous'/'ethnic'/'native'; as if there were no other grounds for an African identity than the 'indigenous' and the 'ethnic'" (2021, 78). Against this consignment I have tried to put multiple epistemic traditions into play and, by translating across them, trace the outlines of a shared "Euro-African" socialism, an exploration of the problems and possibilities of coexistence. If the decolonial project is often predicated on difference as "that which separates and cuts off one cultural or historical entity from another," then the challenge is to "understand difference not as a secessionist gesture, but as a particular fold or twist in the undulating fabric of the universe" (Mbembe 2021, 79–80).

I have found it compelling to think this fold, this interrupting "third" that both cuts and connects, through the concept of mode. I leave it to the reader to determine if, on plugging it in, anything comes through. Latour (2012) once cagily dismissed mode as a "banal and quasi-ecological expression," a sort of typological stamp that can be applied to production or existence. But I confess I find something intuitively appealing about the concept, in part because of the way it can also evoke action, living patterns of movement and doing. François Laplantine (2016), for example, calls for a "modal anthropology" derived from Mauss (1973), who in his seminal lecture on "techniques of the body" attended to ordinary acts like eating, drinking, or dancing—"those fundamental fashions that can be called the modes of life, the model, the tonus" (quoted in Howes 2016, x). The idea came to him in a flash while recovering from an illness in a New York hospital. He recognized that the nurses were walking in a style he had noticed at the cinema—and later noticed that Parisian women too were modifying their gaits in accordance with what they saw on screen. Such styles are noticeable by virtue of the way they mediate different distributions of energetic interference and intensity within an ecology: a storm system swells or ebbs, animals verge on fight or flight (Massumi 2017, 79); epidemics surge or die down (Nguyen 2017); the public grows restive or falls quiet. Modes actualize a virtual possibility space, distributing stress across certain regions of the emergent relation and not others.

An important point here is that for living actors at least, modes of doing can involve a kind of ethical and political attunement, the ability to take in signals from the world around you and respond accordingly. It might involve making room on a crowded bus or casting about for the right word with the right feel (Das 2015, 91; n.d., 2; Connolly 2013, 145), or knowing when not to push your luck, or being able to sense how others respond to your own signs and signals. These are "teleodynamic" adjustments (Connolly 2013, 82), minor and major, through which actors can stay aligned to the world and to each other.

One basic premise underlying the critical discourse of the Anthropocene is that "we" humans are receiving certain atmospheric and stratigraphic signals from the planet that hosts us. For those of us paying attention, such signals can help us extrapolate how that hosting will—or will not—continue to play out (Kohn 2013; Kockelman 2016; Khan 2019). The critical "turn" to infrastructure is, in its own way, an index of this historical moment. As Paul Edwards (2003, 188) notes, infrastructures are associated with the "artificial environments" that concretize (often literally) the modernist settlement that separates nature from culture, delivering "systemic, society-wide control over the variability inherent in the natural environment." And yet the last few postsocialist decades have been marked not by the patient, incremental, and path-dependent refinement of large technical systems, but by global infrastructural mania: wild swings between excess and starvation. The New Orleans levees and Minnesota bridge collapse of the mid-2000s were fitting tributes to the sustained neglect and underinvestment in public goods, as are the leaded pipes and paint that continue to poison children in cities like Flint and Baltimore. Meanwhile, as Euro-American governments pursued an austerity politics up to and through the 2008 economic crisis, BRIC countries such as China and India were investing in massive infrastructure projects as a form of economic stimulus, often leading to surreal scenes of ghostly subway systems and shopping complexes awaiting life to develop around them (Harvey 2012, 61; Johnson 2013). Infrastructures are "media of mutuality" (Rutherford 2016, 70), and their growth and decay foreground the increasingly up-for-grabs nature of collective interdependence. The Julius Nyerere Hydropower Station is an exemplar of this more-than-human distributive politics: it represents a shift in which the flows that nourish tourists, Tanzanians, locals, animals, soils, viruses, the state, and much else will be massively rearranged. The question is which parties will tolerate this rearrangement, and to what extent.

This book has suggested that, against this background, the ethico-political question of how actors remain within emergent form is one important direction in the anthropology of infrastructure. Dar es Salaam's postcolonial power grid has never been part of a seamlessly functional artificial environment. But that very messiness foregrounds the reality that state officials, consumers, utility workers, and "parasitic" *vishoka* are woven together in an interdependent condition, each actively giving and taking from the same central system. This system is in principle coextensive with a postcolonial and African socialist nation that has invested much of its self-understanding in becoming, like a networked infrastructure itself, both "modern" (i.e., industrial, technological) and collective (i.e., interconnective, more than just an aggregate of individuals, ethnicities, or villages). In hooking up to the grid, Tanzanians not only receive electricity, but each other.

And this takes work. In the postsocialist era, the Tanzanian state has proven itself aware that every tariff raise, energy scandal, and power interruption is one more sign of its moral abdication, and has worked to varying degrees to forestall it. Conversely, urban residents weigh the temptations of nonpayment, surreptitious reconnections, or outright bypass against what they infer utility workers will tolerate or, given their own straitened circumstances, facilitate. In short, along nested scales of Tanzanian life—that of the citizenry and state, individual consumer and local utility branch—actors tack back and forth between their own needs and that of the network as a whole. However provisionally they do so, they calibrate the one to fit the other so that both might continue unfolding into the future. This "modal reasoning" might hold broader lessons for theorizing the demands of coexistence. Dar es Salaam's nickname *Bongo* (brains) is a fact often explained from an actor-centered perspective—one needs one's wits to survive. But insofar as survival is ultimately a collective project, perhaps it is the city itself that is doing the thinking.

INTRODUCTION

1 As Tim Ingold (2015, 11) points out, contract should properly refer to an inter-weaving that is evoked in its etymology of *com* (together) + *trahere* (to pull or draw).

2 On Serres's deceptively "acritical" philosophical disposition, see Latour (1987).

3 For a useful critique, see Newell (2006, 182).

4 For a powerful analysis of the ways material flows substantiate social worlds, see Claudia Gastrow's (2017) ethnography of "cement citizenship" in Luanda.

5 Daniel Mains (2012) observed a similar phenomenon in the comparison between privately contracted hydroelectric dam construction and privately contracted road construction in urban Ethiopia. The latter acquired popular buy-in and sacrifice, in part for the ways in which the emergent *effect* of these private contracts was, manifestly, collective development—roads that all could use. The former simply produced blackouts.

6 Actor-network theory tends to imply an "auto-limitlessness" that can only be resolved by some performative act of "cutting the network" (Strathern 1996). Moreover, if a network is simply the sum of the relations that compose it at any given moment, then one is reduced to the untenable philosophical position of "actualism," in which every accretion produces an ontologically new entity (Harman 2010, 128).

7 This quintessentially Melanesianist example finds its Africanist counterpart in the "spheres of exchange" of late colonial Tiv society, whereby rights-in-people (i.e., marriage) is the paramount (we might say infrastructural) sphere since it does not deliver any particular use-value but rather the very possibility of future use-value (Bohannan 1955).

8 As John Durham Peters (2015, 14) remarks, in phatic communication one has "nothing to say and everything to mean." See also Stevenson (2017).

9 Charles Zuckerman (2016) helpfully distinguishes between phatic *contact* qua the holding of psychological attention through a technical medium and phatic *communion* as a more specific instance of "positive affiliative sociality" (297) via the exchange of redundant (i.e., channel-foregrounding) messages that Malinowski seemed to privilege, warning that the former does not always lead to the latter. Rather, culturally specific ways of establishing, suspending, repairing, or switching communicative channels are social acts that can signal whether one is dealing with a friend, enemy, rival, superior, or ex (Gershon 2010; Lemon 2018). Indeed, sometimes cutting off (technical) contact preserves the (social) connection (Degani 2021).

10 In a similar vein, geographers Maria Kaika and Erik Swyngedouw (2000, 121) characterize infrastructural networks such as "water towers, dams, pumping stations, power plants, gas stations" as an "urban dowry."

11 "Tanesco Power for a Changing Society," *The Standard Tanzania*, 1968.

12 Povinelli (2011b, 7) tilts it toward the latter when she suggests we think of these infelicities as the presence of other worlds in formation: "no world is actually one world. The feeling that one lives in the best condition of the world unveils the intuition that there is always more than one world in the world at any one time. The very fact of Malinowski's presence, and his own argument that for the Trobrianders there were worlds within worlds, testifies to this claim. The material heterogeneity within any one sphere, and passing between any two spheres, allows new worlds to emerge and new networks to be added. This heterogeneity emerges in part because of the excesses and deficits arising from incommensurate and often competing interests within any given social space."

13 As Kockelman suggests, "as *The Parasite* is to *Leviathan*, so actor-network theory is to classical sociology" (2017, 36).

14 For a lucid reflection on boredom as the dominant structure of feeling under midcentury Euro-American capitalism, as well as a lively analysis of nineteenth-century "misery" and twenty-first century "anxiety," see The Institute for Precarious Consciousness (2014).

15 For a description of a variation on the theme of roadblock shakedowns in Dar es Salaam, see Tripp (1997, 1). For a contemporary example from Latin America, see Lyon (2020, 1).

16 An interest in modifications not only resonates with a pragmatist tradition that encompasses figures as diverse as Martin Heidegger and the American Transcendentalists, with their interest in dwellings and tools, but the ordinary language philosophy of John L. Austin and Ludwig Wittgenstein, with their interest in the way meaning emerges through inhabitation. For example, drawing on Stanley Cavell, Veena Das reflects on the ways words are projected into new situations, such that there is no essential meaning but rather a sense of what is alive in the sentence. Speakers give words different accents and stresses. To the degree that others receive those words, they are joined up together in the trajectory of its emergent character. But this projection

depends on a feel for what we might call a word's topological invariance or "inner constancy"—the way its meaning can be rotated, folded, or stretched. That we can speak of "feeding the meter" but not, generally speaking, "feeding your love," suggests that understanding a word lies not in describing what it is, but in the different ways it can and can't move within a given form of life (Das 2014, 285). Note that Das's quintessentially Austinian ordinary language example, like Kockelman's spoon, depends on the use of modal auxiliary verbs.

CHAPTER ONE. EMERGENCY POWER

Epigraph: Nimi Mweta, "Anatomy of the Power Crisis: Who Didn't Do What, When," *The Citizen*, July 17, 2011.

1 Here and throughout, the currency referred to is the Tanzanian shilling. At the time of my fieldwork, the average exchange rate was 1600 shillings to US$1.

2 Tanesco also operates eighteen, mostly diesel-powered mini-grids unconnected to the national grid, servicing rural towns or district centers (see Perez 2018). Moreover, since 2010 a number of mostly hydropower SPPs have come online thanks in part to an updated regulatory framework and growing interest in renewable energy investment. Some feed into the national grids, while others extend or feed into the country's mini-grids. The 5 MW Tulila power plant, for example, is run by the Benedictine Sisters of St. Agnes and feeds into the national grid to power the Songea municipality and neighboring villages (Chipole 2021). The 4 MW Mwenga hydropower station and mini-grid service villagers in the rural Njombe and, through an interconnection to the national grid, provide backup power to the area's Mufundi Tea and Coffee factory (USAID 2018).

3 SPPs are defined as private producers with a generation capacity of less than 10 MW.

4 See, e.g., the introductory comments in: The World Bank, "Second Power and Gas Sector Development Policy Operation," February 26, 2014, http://www-wds.worldbank.org/external/default/WDSContentServer/WDSP/IB/2014/03/05/000442464_20140305111847/Rendered/INDEX/840280PGD0P145010B0x38 2156B00OUO090.txt.

5 When President Obama visited the Ubungo power plant in Tanzania in 2013, he participated in a staged photo-op with the Sockett, a soccer ball that stores a small electricity charge. Thirty minutes of play is capable of powering a small LED lamp, though the ball was priced at $99 at the time, while a solar lamp was $10. See Aroon Preeti, "Sorry, Obama, Soccer Balls Won't Bring Progress to Africa," *Foreign Policy*, July 2, 2013, https://foreignpolicy.com/2013/07/02/sorry-obama-soccer-balls-wont-bring-progress-to-africa/.

6 See Ferguson (1994, 18) on the "counter-intentionality of structural production."

7 These include coal (Mitchel 2011), oil (Rogers 2014), solar (Cross and Murray 2018; Günel 2019), methane (Chalfin 2020), wind (Boyer 2019; Howe 2019), and (small scale) hydropower (Whitington 2019).

8 Planners and international development experts, for example, hoped that the socioeconomic development generated by the 1320 MW Kariba Dam in the Central Africa Federation (1953–1963, comprising what is now independent Zambia and Zimbabwe) would unite to "create social cohesion among the Federation's three territories, with their vastly different backgrounds and internal divisions, reconciling 'African' and 'European,' settler and expatriate, rural and urban factions" (Tischler 2014, 1048).

9 No less a Cold Warrior than Robert McNamara was a key supporter of *ujamaa* during his tenure as president of the World Bank in the 1970s, seeing it as aligning with the Bank's rural antipoverty campaigns and allocating significant funds to its projects (Delehanty 2020).

10 The OPEC price hikes of 1979 raised expenditures on fuel imports (from 8 percent to 23 percent in 1982); a withdrawal from the East African community meant it could no longer pool costs for upkeep of regional infrastructure, and a series of recurring droughts all reduced agricultural production. These shocks sparked a vicious circle.

11 These included the Ministry of Finance, with its close relationship to donors, as well as the Ministry of Agriculture, who had long dealt with smallholder dispossession via state marketing boards and villagization (Lofchie 2014, 147).

12 In fact, as David Harvey (2005, 66) notes, neoliberal theorists are often "profoundly suspicious of democracy . . . [and] tend to favour governance by experts and elites." In Tanzania this was exactly the case. The original reformists opposed multipartyism, fearing it would exacerbate the instability caused by structural adjustment, just as international policymakers in Russia and elsewhere promoted market "shock therapy" since "the short-term pain of stabilization and liberalization might knock reformers out of office before structural reforms were complete" (Collier 2012, 147).

13 Jenerali Ulimwengu, "As the Race to Succeed Kikwete Hots up, Lowassa Is the Man to Watch," *The East African*, accessed February 19, 2021, https://www .theeastafrican.co.ke/tea/oped/comment/as-the-race-to-succeed-kikwete-hots -up-lowassa-is-the-man-to-watch-1322888.

14 In 1992, a Joint UNDP/World Bank Energy Sector Management Assistance Programme survey reported that 20 percent of all power generated was lost due to technical factors (11 percent)—that is, power lost as electricity is converted to heat through inefficiencies in transmission and distribution—and "nontechnical" factors (9 percent) from power consumed but not billed or paid for. It also offered a close look at the technical procedures of consumer billing and payment, pointing to the persistence of widespread billing irregularities, lack of enforcement, and degraded physical infrastructure at the distribution level (Hay 1998).

15 This includes Mercados, who undertook the cost-of-service study mentioned in the chapter's introduction and who provided plans for Tanzania's power sector unbundling in the 1990s.

16 Kihansi is also symbolically relevant to the new era, as it marks the last of the technical and hydropower-oriented Scandinavian aid package that had

characterized their donor activities since the 1970s. In the 1990s, Norway and Sweden's priorities then switched to institutional policy reforms.

17 Ostensibly, this was due to lack of funds for the import duty on fuel, an import duty that the government refused to waive. As Cooksey (2002, 52) relates: "Conspiracy theorists in Dar es Salaam [were] convinced that this example of gross mismanagement [was] not accidental, but orchestrated by IPTL and its local supporters to sell their proposal to the public."

18 As Kelsall (2002, 606) observes: "President Mkapa, it seems, would have found it difficult to secure the presidential re-nomination in 2000 without [them]. Moreover, in a context of multiparty democracy, the party finds these individuals useful in securing re-election. Thus the 'old guard' has continued to exercise influence behind the scenes throughout the Third Phase government, and more recently some of its members have found their way back into Cabinet. While the president's own integrity is rarely called into question, the stance of several members of the government with respect to reform is now ambiguous."

19 Much as Nyerere and the World Bank cooperated in the rollout of *ujamaa* policies in the 1970s (see chapter 1, note 9). Indeed, there is a remarkable detail of historical continuity in that it was former World Bank president Robert McNamara who, heading Transparency International in the 1990s, commissioned Cooksey to investigate IPTL (personal communication; see also Cooksey 2002, 50).

20 In 2005, Songas's capacity charge was $3.5 million/month (adjusted for inflation), but this was inclusive of the entire gas processing and transport infrastructure, not just electricity (Gratwick, Ghanadan, and Eberhard 2006, 47).

21 In 2005, these IPPs took 69 percent of Tanesco's total revenue and provided only 45 percent of generation. IPTL alone accounted for 62 percent of IPP payments but only 37 percent of IPP generation. In 2006 renewed drought worsened these numbers, with IPPs accounting for 55 percent of all generation and a whopping 96 percent of revenue (Ghanadan and Eberhard 2007, 18). It should be noted that despite these adverse conditions, the structure of management bonuses was based on revenue collected that did not include IPP charges; they were paid handsomely.

22 In 2004, it transferred 125 billion shillings (or 1 percent of GDP) to Tanesco "to cover the deterioration of its finances resulting from drought, oil prices and the financial impact of past investment decisions taken by the government" (Matiku, Mbwambo, and Kimeme 2011, 201).

23 Nimi Mweta, "Anatomy of the Power Crisis."

24 In the following account, I rely primarily on a parliamentary select committee report that was commissioned to investigate the Richmond scandal (Bunge la Tanzania 2007).

25 "Who Really Owns Dowans?," *The East African*, January 31, 2011.

26 The Tanzania and New York branches of Citibank refused to open the letter of credit in cooperation with Tanzania's CRDB bank, citing insufficient time to conduct due diligence, whereas HSBC bank eventually agreed.

27 I take this notion from Slavoj Zizek's (2005, 52) discussion of the *salto mortale*, or leap of faith, that occurs in the realization of exchange value: "the temporality here is that of the *futur antérieur*: value 'is' not immediately, it only 'will have been,' it is retroactively actualized, performatively enacted. In production, value is generated 'in itself,' while only through the completed circulation process does it become 'for itself.'" On the performativity of the bluff in African capitalist modernity, see Newell (2012).

28 Though it is beyond the scope of this chapter to address, Tanzania also navigated an IPTL redux scandal involving over US $100 million. With the help of senior government officials, IPTL's shares were fraudulently transferred to another company, Pan African Power (PAP), and disputed funds held in a government escrow account (Tanesco alleged IPTL had been overcharging and sought recourse through commercial arbitration) were then transferred to PAP. In many ways the "Tegeta Escrow" scandal recapitulates the dynamics of emergency power seen in the Richmond and original IPTL sagas. See Degani (2017).

29 I refer to Weber's famous image of the "light cloak" of care for worldly goods hardening to an "iron cage" ([1930] 2005, 123). Peter Baehr (2002) suggests the latter would be better translated as "steel shell," an image with vitalist overtones that are perhaps resonant with the social effects of energy use.

30 The same faction that sponsored the parliamentary select committee investigation. The Chair, Harrison Mwayakembe, and his ally Samuel Sitta, were prominent CCM members that had allegedly been sidelined by Lowassa factions. "Ministry of Power Struggles," *Africa Confidential*, October 8, 2011.

31 Symbion is itself an American example of the kind of post–Cold War oligarchic formation discussed above. Its board was chaired by former Ambassador Joseph Wilson, who was also director of Jarch Capital involved in Southern Sudan. Wilson held a close relationship to Secretary of State Hillary Clinton and wrote to her mentioning that Symbion would be bidding on the Millennium Challenge Account, which she chaired. Symbion also won government contracts for work in Iraq and Afghanistan. See Isaac Arnsdorf, "Clinton Gave Nod to Ex-Ambassador Wilson's Company for Contract," *Politico*, January 8, 2016. https://www.politico.com/story/2016/01/hillary-clinton-joe-wilson -symbion-217497.

32 "Remarks by President Obama at Ubungo Symbion Power Plant," The White House Office of the Press Secretary, accessed March 3, 2015, http://www .whitehouse.gov/the-press-office/2013/07/02/remarks-president-obama-ubungo -symbion-power-plant.

33 "US$5 Million a Month Loss for Tanesco," November 27, 2007, http://www.esi -africa.com/us-5-million-a-month-loss-for-tanesco/.

34 "Tanesco Bosses Fired over Sh. 280 Billion Loan Scam," *The Citizen*, March 7, 2010. Rashidi is probably more famously known for his involvement in the "radar scandal" involving the overinflated sale of an air-traffic controller system to the Tanzanian government.

35 Mike Mande, "Probe Unearths Flawed Tenders, Deals at Tanesco," *The East African*, November 3, 2012.

36 Florian Kaijage, "Power Crisis Looms Amid Battle for Tanesco's Soul," *The Citizen* (Tanzania), July 7, 2012.

37 "Mgao Feki wa Umeme," *Raia Mwema*, July 18, 2012.

38 "Tanesco Board Sacks Mhando," *The Guardian* (Tanzania), November 1, 2012.

39 Nimi Mweta, "Tanesco Pressure Method, Power Cuts Back with a Vengeance," *The Citizen* (Tanzania), November 18, 2012.

40 Nimi Mweta, "The Muhongo Effect—Is It Durable as Circumstance, or as Style?," *The Citizen* (Tanzania), August 6, 2012.

41 Mweta, "Anatomy of the Power Crisis."

CHAPTER TWO. THE FLICKERING TORCH

Epigraph: Julius Nyerere, "A Candle on Kilimanjaro," address to the Tanganyika Legislative Council, October 22, 1959 (quoted in Mwakikagile 2006, 104).

1 Mlagiri Kopoka, "Statistics That Mean Nothing," *The Citizen*, January 28, 2013.

2 In Kockelman's terms, these sorts of inferences are relatively deductive and remain at a relatively shallow level of "ontological transformativity." Inferences can go "deeper," via induction and abduction (2017, 180–83).

3 In November 2005, for example, a third of Dar es Salaam's population experienced load shedding for nearly a month because two transformers at the Ilala substation broke down. Hence the area could not physically receive enough power for demand, prompting first a blackout followed by a managed schedule of power rationing as a new transmission line was constructed. "Normal Power Supply Returns to Dar es Salaam," Xinhua News Agency, November 22, 2005.

4 "Make Tanesco Gesture Lasting," *The Citizen*, December 26, 2012.

5 "Tanzania Extends Power Cuts," *Water Power and Dam Construction*, March 23, 2006.

6 "Power Rationing Ends in Tanzania as Dams Fill Up," BBC *Monitoring Africa*, April 5, 2006.

7 "Power Rationing to Continue in Tanzania, Says President," BBC *Monitoring Africa*, May 31, 2006; "Blackout to Affect Tanzanian Fans Watching World Cup," *Xinhua General News Service*, June 8, 2006; "Tanzania: Dar Must Move On From Hydro," *Africa News*, June 13, 2006.

8 "Tanzania: Power Cuts to Worsen Following Mechanical Breakdown," BBC *Monitoring Africa*, August 26, 2006.

9 "Power Crisis in Tanzania to Worsen as Dam Closes," *Xinhua General News Service*, October 18, 2006.

10 "Tanzania Lengthens Power Blackout Time," *Xinhua General News Service*, November 19, 2006.

11 "Tanzania Ponders Power Tariff Hike as Supplies Ease," *Xinhua General News Service*, December 7, 2006.

12 "Tanzania: State Electricity Firm Ends Power Rationing," BBC *Monitoring Africa*, December 29, 2006.

13 These included regular weekend shutdowns in Arusha due to improvement of the transmission system, a ninety-five-hour-long blackout throughout the entire Shinyanga region. September and October of 2008 saw a few weeklong ten-hour rationing periods, thanks to a breakdown in the Songas turbines and a blown transformer at the Ubungo power plant. Other breakdowns in the transmission system created twelve-hour-long outages in parts of Dar es Salaam for three days in early February of 2009. "Tanzania: Power Cuts Still Order of the Day," *Africa News*, February 24, 2007; "Power Utility Suffers Loss over Blackout in Northwest," BBC *Monitoring Africa*, August 29, 2008; "Tanzania: Regions to Go without Power Nine Hours a Day," *Africa News*, October 8, 2008; "TANESCO Warns of Power Crisis in Tanzania," Xinhua General News Service, February 4, 2009.

14 "Dar Halts Power Rationing," *Africa News*, October 24, 2009.

15 "Tanzania: Tanesco Allays Fears on Fresh Electricity Cuts," *Africa News*, December 8, 2009.

16 "Tanzania: What Lies Behind Power Cuts in Dar," *Africa News*, December 28, 2009.

17 "So Help Me God," *The Citizen*, November 27, 2010.

18 "Tanesco Issues Nationwide Blackout Alert," *The Citizen*, November 27, 2010.

19 "Power Cuts to Continue through Xmas, New Year," *The Citizen*, December 23, 2010.

20 "Tanesco Says Power Crisis May Ease Soon," *The Citizen*, February 10, 2011.

21 "Symbion Power, Tanesco to Provide Electricity to Network," *Tenders Info*, June 10, 2011.

22 "Tanesco Announces 12-Hour Power Cuts," *The Citizen*, June 22, 2011; "Why Power Rationing Is Needless Suffering," *The Citizen*, June 23, 2011.

23 "The Snafu which the Nation's Energy Sector Has Become," *The Citizen*, July 27, 2011.

24 "Government Terms Power Rationing as 'National Disaster,'" *The Guardian*, June 27, 2011.

25 "Government Terms Power Rationing as 'National Disaster.'"

26 "Muster Resources to Tackle Power Woes," *The Citizen*, August 1, 2011.

27 "No Lasting Solution Yet to Power Crisis," *The Citizen*, March 4, 2011.

28 "Power Cuts Affect the Price of Sugar," *The Citizen*, February 27, 2011.

29 "VETA—Will Reforms Usher in New Era?," *The Citizen*, April 5, 2011.

30 "Dar Falls 12 Places in Global Tourism Index," *The Citizen*, March 30, 2011; "IMF Set to Lower Forecast," *The Citizen*, March 9, 2011.

31 "Small Businesses Feel the Impact of Power Cuts," *The Citizen*, May 23, 2011.

32 "Electricity Has to Be a Fundamental Human Right," *The Daily News*, October 24, 2015.

33 "Candid Talk—Why I Am in 'Love' with Tanesco," *The Citizen*, June 26, 2011.

34 "Fury over Bomb Blast," *The Citizen*, February 17, 2011.

35 "How the Blackouts Have Changed our Lifestyles," *The Citizen*, February 24, 2011.

36 "With Such Meek Citizens, Who Needs Riot Police?," *The Citizen*, January 11, 2011.

37 The invocation of Shylock demanding his pound of flesh was the same trope Julius Nyerere memorably used in the 1980s when describing structural adjustment conditionalities imposed by IMF loans (see Tripp 1997, 79).

38 "Mgao wa Umeme Usio na Kikomo: Watanzania Wamezowea Kunyanyaswa?," June 23, 2011, http://www.chahali.com/2011/06/mgao-wa-umeme-usio-na-kikomo-watanzania.html.

39 Diadie Ba, "Protests Erupt in Senegal over Worsening Power Cuts," Reuters, June 28, 2011, https://www.reuters.com/article/ozatp-senegal-protests-20110628-idAFJOE75R0ID20110628.

40 "Blame It All on the Weather, It Can't Talk Back," *The Nation* (Nairobi), May 8, 2000.

41 Nkwazi Mhango, "We Laughed at Them: They Are Laughing at Us!," *Mpayukaji*, November 10, 2007, https://mpayukaji.blogspot.com/2007/11/we-laughed-at-themtheyre-laughing-at-us.html.

42 It is obvious but important to state that the above description in no way implies that early Tanzanian nationalism was that of total unity and consensus. Ethico-political contestations swirled around and through TANU's consolidation of hegemony, and Tanzanians of all stripes resisted state control through various combinations of exit and voice (see Maddox and Giblin 2005).

43 Indeed sometimes literally, as in the case of the annual Uhuru torch race, where the ever-lit monument is carried around different parts of the country before it is planted atop Kilimanjaro.

44 January Makamba, "Politics and Power Cuts in Tanzania," *Politics, Society and Things*, June 29, 2011, https://taifaletu.blogspot.com/2011/06/politics-and-power-cuts-in-tanzania.html.

45 The play of presence and absence of recalling the structure of Eastern Bloc jokes about planned economies—"we pretend to work and they pretend to pay us."

46 "Sitta—Punish Those behind Power Crisis," *The Citizen*, June 17, 2011.

47 It must be said that CCM did not, much to their chagrin, have a monopoly on Nyerere. In the 2010 Presidential election CHADEMA candidate Wilibraod Slaa also ran on the CHADEMA ticket with the slogan *Nyerere Hadi Slaa* (From Nyerere to Slaa) (Fouéré 2014, 47). Supporters of CCM objected to this appropriation of their party's founder, but after all, it was Nyerere himself who had late in life advocated for multiparty elections, showing little patience for filial obedience to a party per se.

48 S. M. Mgaya, "Practical Methods of Tackling Power Crisis in Tanzania," speech delivered at *Meeting to Debate the Electricity Problem in Tanzania*, July 30, 2011.

49 Makamba, "Politics and Power Cuts."

50 "Muster Resources to Tackle Power Woes."

51 "Occupy Arusha, Dar, Mwanza, Mbeya . . . 'Imetosha,'" *Occupy Arusha*, October 21, 2011, http://occupydar.blogspot.com/2011/10/occupy-dar-october-21-2011

-imetosha.html; Elsie Eyakuze, "A New World Order: Occupy Yourself First," *The Mikocheni Report*, October 26, 2011, https://mikochenireport.blogspot.com /2011/10/new-world-order-occupy-yourself-first.html.

52 McGovern (2010, 56–57) offers a similar dialectic of "gerontocratic hierarchy" and "entrepreneurial capture," as I discuss in chapter 4. See also Steve Feierman's discussion (1990, 76–85) of *kubana shi* (harming the land) and *kuzifya shi* (healing the land) in precolonial Shambaai.

53 "Ni Hamsini, Ni Hamsini: Tanzania at 50, What Do We Have to Show for It?," *The Daily News* (Tanzania), December 7, 2011.

54 This is Wittgenstein's description of mindlessness (quoted in Das 2015, 61).

55 Mhango, "We Laughed at Them."

56 Mlagiri Kopoka, "Bongo: Land of Broken Promises," *The Citizen*, February 28, 2012.

57 "Tanzania's Jakaya Kikwete Denies Steven Ulimboka's 'Torture,'" BBC News Africa, July 2, 2012, http://www.bbc.co.uk/news/world-africa-18671315.

58 "Why Govt Delays Threaten Taneco 408bn Projects," *The Citizen*, March 18, 2012.

59 "Power Cuts to Continue in Dar es Salaam," *The Citizen*, March 12, 2012.

60 "Why Govt Delays Threaten Taneco 408bn Projects."

61 "Tanesco—Rationing Is Back," *The Citizen*, March 27, 2012; "Rationing Not Linked to Tariff Rise Proposal, Says Tanesco," *The Citizen*, November 19, 2013.

62 "10 Reasons Why Power Rationing Makes Little Sense," *VijanaFM*, November 8, 2013, http://www.vijana.fm/2013/11/28/10-reasons-why-power-rationing -makes-little-sense.

63 "Tanesco Pressure Method, Power Cuts, Back with a Vengeance," *The Citizen*, November 18, 2012.

64 "Akaunti ya Kigogo Tanesco Yazuiwa," *Raia Mwema*, July 18, 2012.

65 "Tanesco Yanafanfanua Kuhusu Vishoka," *Majira*, June 7, 2011.

66 My translation. Samwel Munro, Twitter post, August, 22, 2012, http://twitter .com/murosam.

67 https://twitter.com/TANESCO_.

68 Tanesco Twitter post, February 28, 2013, https://twitter.com/TANESCO_ /status/307373401536880640.

69 Tanesco Twitter post, March 1, 2013, https://twitter.com/TANESCO_/status /307431322471776258.

70 Tanesco Twitter post, February 28, 2013, https://twitter.com/TANESCO_ /status/307379028992589824.

71 Tanesco Twitter post, March 1, 2013, https://twitter.com/TANESCO_/status /307373401536880640.

72 Serres's parasite can also be included in this series. By making relations nothing but parasites all the way down, that is, "composed of nothing but an endless stretch of endlessly swappable and scalable nodes and edges . . . Serres did away with most forms of traceable identity . . . and grounded locality" (Kockelman 2017, 47). Actor-network theory would seem to celebrate this fact.

But entropy, for instance, is a key embodiment of the parasitic relation for Serres, as is noise. In their own ways, runaway climate change and unchecked pandemics embody the malevolence of being caught within a network of continuous rhizomatic mutation.

73 "Tanzanian Police Clash with Protesters over Arrested Cleric," Reuters, November 2, 2012, https://af.reuters.com/article/topNews/idAFJOE8A101X20121102.

74 Elsie Eyakuze, "Anxiety Attacks," *The Mikocheni Report: A Life in Dar es Salaam*, accessed August 23, 2021, http://mikochenireport.blogspot.com/2012/10/anxiety-attacks.html.

75 See Clara Han's account of a similar tactic of "cutting off the lights" (2012, 1) in a poor Chilean neighborhood, in anticipation of a confrontation with security forces on the anniversary of the 1973 assassination of socialist president Salvadore Allende and subsequent coup.

76 "Nation Set for Power Boom by 2015," *The Citizen*, November 9, 2012.

77 Elsie Eyakuze, "Instead of Live Coverage of Election Rallies, We Get the Poetry of Dar under a Full Moon," *The East African*, October 10, 2015.

78 "We're Switching Off All Hydro-Power Plants: Minister," *The Citizen*, October 12, 2015.

79 "Kinyerezi Gas Plants On, But Power Blues Persist," *The Citizen*, September 27, 2015.

80 "JK Hurt He Ends Term with Power Rationing," *The Citizen*, October 14, 2015.

81 "Lowassa Vows to Clean Up Tanesco, Lazy Staff Warned," *The Citizen*, October 7, 2015.

82 "Power Failure Cuts Short Lowassa's Tunduma Rally," *The Citizen*, October 18, 2015.

83 While there is not to my knowledge any scholarship on the use of power cuts to *depress* electoral participation, there is a burgeoning political science literature on the increase of electricity supply to constituents as a campaign tool in the runup to elections (e.g., Min and Golden 2014; Imami et al. 2020).

84 Eyakuze, "Instead of Live Coverage."

CHAPTER THREE. OF METERS AND MODALS

1 The relationship between the anthropologist and research assistant has been subject to insightful critique and reappraisal. For a summary, see Middleton and Cons (2014). See Degani (2021) for an attempt to more thoroughly foreground Thierry's important role in ethnographic knowledge production, particularly during DC patrols.

2 One estimate placed the ratio of Tanesco staff members to customers at this time at 1 to 1,260 (Azorom, AETS [Application européenne de technologies et service] 2011).

3 "Pay Tanesco Debt Lest It Collapses," *The Citizen*, November 15, 2011.

4 This was the response to my household survey and the survey conducted by Ghanadan (2008, 95).

5   A study in Ghana (Dzansi et al. 2018) similarly found that load shedding positively correlated with increased nonpayment.

6   As in the tradition of the *nyumba ndogo* ("little house")—the mistress or second family (Lewinson 2006b).

7   Reliable data about the gender composition of Tanesco's labor force is hard to come by, but some surveys of Tanesco's workforce and organizational culture produced over the 2010s lead me to estimate that anywhere from 20–40 percent of employees in Tanesco's administrative and technical departments were women (with the exception of service line installation crews, which were exclusively male) (see Mwinami 2014, 44; Mbilinyi 2013; Lukumai 2006, 28; Due and Temu 2002, 331). Beyond this rough assessment of numbers, I find it difficult to generalize about the ways gender impacted the way employees carried out their work, though I make some observations below.

8   In a faint but strange historical echo of Sabina's encounter, colonial-era rumors of *mumiani* came to a head in 1959 when rumors of a young girl's abduction by vampiric state agents prompted a large-scale riot in the "African" settlement of Buguruni (Brennan 2008).

9   He comments: "When we went to Singapore, we could see their satellite towns, their ring-roads, their skyscrapers and their decentralised services, and it's working very nicely there." Joe Boyle, "Dar es Salaam: Africa's Next Megacity?," *BBC News*, July 30, 2012, http://www.bbc.co.uk/news/magazine-18655647. This uncannily echoes the pilgrimage east that Nyerere took to China in the 1970s, where he too was impressed with state-directed development, and inspired his compulsory villagization. Even more could be said about the symbolic landscapes of China versus Singapore, the former indexing the vast magnitude of socialist industrialism, Singapore the tidy, enclaved, post-industrial city state.

10 The idea that the state is both "autocratically over-centralized," yet unable to "capture the peasantry" for the requirements of development structured 1980s debates about etatization (Ferguson 1994, 172).

11 Perhaps the most spectacular case of violence against Tanesco workers took place in February of 2010. A Bishop Kakobe of the large and wealthy Full Gospel Bible Fellowship Church was ordered by Tanesco to remove a colorful billboard over his chapel to make way for a high voltage transmission line. Kakobe rallied his followers, who donned yellow shirts that said "Tanesco, Fear God" and stood guard on the premises for three days. When Tanesco surveyors arrived to take measurements, they were attacked and beaten. While in many ways this episode is an example of more general struggles over urban development and land use, it in part gained traction for the way that Kakobe capitalized on widespread resentment of Tanesco and used a religious idiom to demonize its workers as unnatural interlopers. "How Kakobe Lost the Battle to Tanesco," *The Citizen*, March 29, 2010.

12 Beyond being a social dependent, it was unclear what the young girl's relationship was to the absent adults of the house—hired domestic help, distant kin or immediate relative.

13 See Degani (2021) for a fuller account.

14 This is not to say that my specific position as an ethnographer did not have its own unique inflections. Understanding Kiswahili on rounds set me apart from other *wazungu* (whites), leading to some surprising and funny shared moments (e.g., Degani 2018, 492). People also gawked and laughed at the spectacle of an *mzungu* riding in the back of a utility pickup truck with DC and RPU teams or winding up wire with installation crews. Of the latter, one spectator mischievously shouted to his friends that he wholeheartedly approved—these *wazungu* were willing to work hard, no matter what!

15 Most obviously expressed in the figure of a witch. See Newell (2007) for an excellent discussion.

16 By 2000, for example, civil servant salaries provided on average one fifth of the purchasing power to equivalent salaries in the 1970s (Mutahaba 2005, 3).

17 "Tahadhari Kwa Wateja Juu ya Kuibuka kwa Wimbi la 'Vishoka—Matapeli' Maeneo Mbalimbali Nichini," Umeme Forum, May 2, 2014, http://umemeforum.blogspot.com/2014/05/tangazo.html.

18 "Fake Luku Inspectors," Wapendwa ktk Mtandao, Yahoo Groups, January 22, 2013, https://groups.yahoo.com/neo/groups/wapendwa/conversations/messages/1193.

CHAPTER FOUR. BECOMING INFRASTRUCTURE

1 Cf. Felix Ravaisson's theory (2008) of people's habits as spanning "grace" and "addiction": grace being the ability to move through the world with maximum fluidity such that the world seems like an extension of one's own will; addiction inversely referring to the subject's maximum immobility such that he or she seems to be its derivative expression. See also Kockelman's (2016) discussion of grace as achieving and occupying a higher value "gradient."

2 This was one of the problems of rebel ex-combatants in Sierra Leone, for instance, who were consigned to a kind of permanent adolescence after the war, not because they were denied the chance for material advancement but because they forcibly took it. Thus they could be seen as stunted by their experiences in the bush, and unable to fully return and assume the social restraint of adult life. See Bürge (2011).

3 See David Graeber's theory (2001) of "action and reflection."

4 Along these lines, Dua (2019, 150) offers a useful distinction between capture (as in livestock capture, bride capture, and, we can add, *utumwa*) as a "mode of engendering and sustaining sociality" and *captivity* as a state of social death exemplified in the transatlantic slave trade. On the capture of "energy feeds" along Africa's resource frontiers, see Degani, Cross, and Chalfin (2020).

5 We might also be reminded of "Wittgenstein's sense of the child who moves about in his or her culture un-seen by the elders and who has to inherit his or her culture as if by theft" (Das 1998, 174).

6 See Doherty (2017) on "disposability."

1   I thank Jim Scott for recommending this novel to me.

2   In July 2011, at the height of Tanesco's power rationing, one columnist published a reminiscence about the socialist-era shortages that drew the parallel directly. During that time, he reflected "Tanzanians started moving around with what in Russia are known as 'perhaps-bags'—and cash. In case one came across items on sale somewhere along one's wanderings, one didn't have to miss buying some only because one wasn't carrying the requisite cash, or a 'perhaps' shopping bag!" "Remember the Days of 'Perhaps' Bags and 'Kaya' Shops? I Still Do," *The Citizen*, July 29, 2011.

3   Hence the founding charter of its anthropology (Verdery 1996) comes in the form of a question: what *was* socialism anyway?

4   For a further discussion of this rhetoric of ambivalent recrimination, see Stroeken (2005).

5   "Tanzanian Opposition Party Challenges Vote Count, Cites Rigging," Reuters, October 28, 2015.

6   "Water Dodgers Face Magufuli's Wrath," *Tanzanian Daily News*, June 22, 2017.

7   "IPTL 'Owners' Finally Charged," *The Citizen*, July 20, 2017.

8   Popular heads of state of the Democratic Republic of Congo, South Africa, Burkina Faso, and Tanzania, respectively. Both Lumumba and Sankara were assassinated.

9   "Ki-mission town" (mission town style) is slang for hustling—literally, at least as one friend explained, of having some sort of mission in the city.

10 Ivana Kottasová, "They Failed Mandatory Tests at School. Then They Were Expelled," CNN Health, October 11, 2018, https://www.cnn.com/2018/10/11/health/tanzania-pregnancy-test-asequals-intl/index.html.

11 Rachael McLellan, "Why Is Once-Peaceful Tanzania Detaining Journalists, Arresting Schoolgirls and Killing Opposition Leaders?," *Washington Post*, November 30, 2018, https://www.washingtonpost.com/news/monkey-cage/wp/2018/11/30/why-is-once-peaceful-tanzania-detaining-journalists-arresting-schoolgirls-and-killing-opposition-leaders/.

12 McLellan, "Why Is Once-Peaceful Tanzania Detaining Journalists?"

13 Simon Allison, "Shot 16 Times, but Still Defiant," *Africa Guardian and Mail*, February 15, 2019, https://mg.co.za/article/2019-02-15-00-shot-16-times-but-still-defiant/.

14 "Coronavirus: John Magufuli Declares Tanzania Free of COVID-19," *BBC News*, June 8, 2020, https://www.bbc.com/news/world-africa-52966016.

15 "Muhimbili and Mloganzila Hospitals Install Steam Inhalation Machines," *The Citizen*, March 4, 2021, https://www.thecitizen.co.tz/tanzania/news/muhimbili-and-mloganzila-hospitals-install-steam-inhalation-machines—3312044.

16 Priya Sippy, "Tanzania's New Leader Is Making Up for Lost Time in the Fight against COVID," Quartz Africa, May 7, 2021, https://qz.com/africa/2006013/tanzania-president-samia-hassan-issues-new-covid-19-restrictions/.

17 Nolan Quin, "Arrests of Tanzanian Opposition Underline Need for Constitutional Reform," August 4, 2021, https://www.cfr.org/blog/arrests-tanzanian-opposition-underline-need-constitutional-reform.

18 Lily Kuo, "Video: Ory Okolloh Explains Why Africa Can't Entrepreneur Itself Out of Its Basic Problems," Quartz Africa, September 15, 2015, https://qz.com/africa/502149/video-ory-okolloh-explains-why-africa-cant-entrepreneur-itself-out-of-its-basic-problems/.

19 Evidenced, for instance in the Silicon Valley hype over Massive Online Open Courses, or MOOCs (Curinga 2016) or in the overengineered wholesale electricity market that helped usher in the 2021 Texas Power crisis (Littlechild and Kiesling 2021).

20 On the high modernist social engineering of the Tennessee Valley Authority, see Scott (2006).

21 "Hotuba ya Magufuli Uzinduzi Stiegler's Gorge," Mtanzania Digital, accessed June 5, 2021, https://www.youtube.com/watch?v=Ovkv-yiMBn8.

22 Over the last decade, ivory poaching has thrived in the Selous Reserve (Kyando et al. 2017). On the economies of poaching (and antipoaching), particularly in central Africa, and the ways in which they weave state and nonstate actors together, see Lombard (2016).

Adwek, George, Shen Boxiong, Paul O. Ndolo, Zachary O. Siagi, Chebet Chepsai-
gutt, Cicilia M. Kemunto, Moses Arowo, John Shimmon, Patrobers Simiyu,
and Abel C. Yabo. 2020. "The Solar Energy Access in Kenya: A Review
Focusing on Pay-As-You-Go Solar Home System." *Environment, Development
and Sustainability* 22 (5): 3897–938.

Ahearne, Robert, and John Childs. 2018. "'National Resources'? The Fragmented
Citizenship of Gas Extraction in Tanzania." *Journal of Eastern African Studies*
12 (4): 696–715.

Akrich, Madeleine. 1992. "The De-scription of Technical Objects." In *Shaping
Technology/Building Society: Studies in Sociotechnical Change*, edited by Wiebe
Bijker and W. & John Law, 205–24. MIT Press.

Amin, Samir. 1978. *Imperialism and Unequal Development*. Harvester Press.

Aminzade, Ronald. 2001. "The Politics of Race and Nation: Citizenship and Afri-
canization in Tanganyika." In *Political Power and Social Theory*, vol. 14, edited
by Diane E. Davis, 53–90. Emerald Group Publishing.

Aminzade, Ronald. 2003. "From Race to Citizenship: The Indigenization Debate
in Post-Socialist Tanzania." *Studies in Comparative International Development* 38
(1): 43–92.

Aminzade, Ronald. 2013. *Race, Nation, and Citizenship in Postcolonial Africa: The Case
of Tanzania*. Cambridge University Press.

Anand, Nikhil. 2012. "Municipal Disconnect: On Abject Water and Its Urban
Infrastructures." *Ethnography* 13 (4): 487–509.

Anand, Nikhil. 2015. "Accretion." Cultural Anthropology website, September 24.
https://culanth.org/fieldsights/715-accretion.

Anand, Nikhil. 2017. *Hydraulic City: Water and the Infrastructures of Citizenship in Mumbai*. Duke University Press.

Anand, Nikhil. 2020. "Consuming Citizenship: Prepaid Meters and the Politics of Technology in Mumbai." *City & Society* 32 (1): 47–70.

Anand, Nikhil, Akhil Gupta, and Hannah Appel, eds. 2018. *The Promise of Infrastructure*. Duke University Press.

Andreasen, Manja Hoppe, and Jytte Agergaard. 2016. "Residential Mobility and Homeownership in Dar es Salaam." *Population and Development Review*, 95–110.

Appel, Hannah. 2019. *The Licit Life of Capitalism: US Oil in Equatorial Guinea*. Duke University Press.

Arendt, Hannah. 2003. *The Portable Hannah Arendt*. Penguin.

Askew, Kelly M. 2002. *Performing the Nation: Swahili Music and Cultural Politics in Tanzania*. University of Chicago Press.

Askew, Kelly M. 2006. "Sung and Unsung: Musical Reflections on Tanzanian Postsocialisms." *Africa* 76 (1): 15–43.

Azarya, Victor, and Naomi Chazan. 1987. "Disengagement from the State in Africa: Reflections on the Experience of Ghana and Guinea." *Comparative Studies in Society and History* 29 (1): 106–31.

Azorom, AETS (Application Européenne de Technologies et Service). 2011. "Loss Reduction Study Tanesco and Zeco: Final Report Executive Summary." Ministry of Finance Millennium Challenge Account Tanzania. Accessed January 24, 2017. http://www.mca-t.go.tz/en/9-news/news/243-electricity-loss-reduction-study-report-executive-summary-v15-243.html.

Baehr, Peter. 2002. "The Iron Cage and the Shell as Hard as Steel: Parsons, Weber, and the Stahlhartes Gehäuse Metaphor in the Protestant Ethic and the Spirit of Capitalism." *History and Theory* 40 (2): 153–69.

Bagachwa, Mboya S. D., and A. Naho. 1995. "Estimating the Second Economy in Tanzania." *World Development* 23 (8): 1387–99. https://doi.org/10.1016/0305-750X(95)00055-H.

Barkan, Joel D. 1994. *Beyond Capitalism vs. Socialism in Kenya and Tanzania*. East African Publishers.

Bayart, Jean-Francois. 1993. *The State in Africa: The Politics of the Belly*. Longmans.

Bayart, Jean-Francois, Stephen Ellis, and Beatrice Hibou. 2001. *The Criminalization of the State in Africa*. Currey.

Bayliss, Kate. 2002. "Privatization and Poverty: The Distributional Impact of Utility Privatization." *Annals of Public and Cooperative Economics* 73 (4): 603–25.

Bayliss, Kate, and Ben Fine. 2007. *Privatization and Alternative Public Sector Reform in Sub-Saharan Africa: Delivering on Electricity and Water*. Palgrave Macmillan.

Bear, Laura, and Nayanika Mathur. 2015. "Introduction: Remaking the Public Good: A New Anthropology of Bureaucracy." *The Cambridge Journal of Anthropology* 33 (1): 18–34.

Beidelman, Thomas O. 1997. "The Cool Knife: Imagery of Gender, Sexuality, and Moral Education in Kaguru Initiation Ritual." Smithsonian Institution Press.

Benjamin, Walter. 2012. *Illuminations: Essays and Reflections*. Schocken Books.

Bennett, Jane. 2005. "The Agency of Assemblages and the North American Black-out." *Public Culture* 17 (3): 445–65.

Berry, Sara. 1989. "Social Institutions and Access to Resources." *Africa* 59 (1): 41–55.

Bertelsen, Bjørn Enge. 2016. "Effervescence and Ephemerality: Popular Urban Uprisings in Mozambique." *Ethnos* 81 (1): 25–52.

Bohannan, Paul. 1955. "Some Principles of Exchange and Investment among the Tiv." *American Anthropologist* 57 (1): 60–70.

Bourdieu, Pierre. 1977. *Outline of a Theory of Practice*. Translated by Richard Nice. Cambridge University Press.

Boyer, Dominic. 2014. "Energopower: An Introduction." *Anthropological Quarterly* 87 (2): 309–33.

Boyer, Dominic. 2015. "Anthropology Electric." *Cultural Anthropology* 30 (4): 531–39.

Boyer, Dominic. 2019. *Energopolitics: Wind and Power in the Anthropocene*. Duke University Press.

Branch, Adam, and Zachariah Mampilly. 2015. *Africa Uprising: Popular Protest and Political Change*. Zed Books.

Brankamp, Hanno. 2015. "#WhatWouldMagufuliDo Sparks New Bout of Tanza-philia." African Arguments, November 30. https://africanarguments.org/2015/11/30/whatwouldmagufulido-sparks-new-bout-of-tanzaphilia/.

Braudel, Ferdinand. 1992. *Civilization and Capitalism, 15th–18th Century*. Vol. 2, *The Wheels of Commerce*. University of California Press.

Brennan, James. 2006a. "Blood Enemies: Exploitation and Urban Citizenship in the Nationalist Political Thought of Tanzania, 1958–75." *The Journal of African History* 47 (3): 389–413.

Brennan, James. 2006a. "Youth, the Tanu Youth League and Managed Vigilan-tism in Dar Es Salaam, Tanzania, 1925–73." *Africa* 76 (2): 221–46.

Brennan, James. 2008. "Destroying Mumiani: Cause, Context, and Violence in Late Colonial Dar es Salaam." *Journal of Eastern African Studies* 2 (1): 95–111.

Brennan, James. 2012. *Taifa: Making Nation and Race in Urban Tanzania*. Ohio University Press.

Briggs, John, and Davis Mwamfupe. 1999. "The Changing Nature of the Peri-Urban Zone in Africa: Evidence from Dar-Es-Salaam, Tanzania." *The Scottish Geographical Magazine* 115 (4): 269–82.

Brownell, Emily. 2014. "Seeing Dirt in Dar Es Salaam: Sanitation, Waste and Citizenship in the Post-Colonial City." In *The Art of Citizenship in African Cities: Infrastructures and Spaces of Belonging*, edited by Momadou Diouf and Rosalind Fredericks, 209–29. Palgrave MacMillan.

Brownell, Emily. 2020. *Gone to Ground: A History of Environment and Infrastructure in Dar es Salaam*. University of Pittsburgh Press.

Buck-Morss, Susan. 1991. *The Dialectics of Seeing: Walter Benjamin and the Arcades Project*. MIT Press.

Bunge la Tanzania. 2007. "Taarifa ya Kamati Teule Iliyoundwa na Bunge la Jam-huri ya Muunagano wa Tanzania Tarehe 13 Novemba, 2007, Kuchunguza Mchakato wa Zabuni ya Uzalishaji Umeme was Dharura Ulioipa Ushindi

Richmond Development Company LLC ya Houston Texas, Marekani, Mwaka 2006." Parliamentary report.

Bürge, Michael. 2011. "Riding the Narrow Tracks of Moral Life: Commercial Motorbike Riders in Makeni, Sierra Leone." *Africa Today* 58 (2): 58–95.

Burton, Andrew. 2007. "The Haven of Peace Purged: Tackling the Undesirable and Unproductive Poor in Dar es Salaam, ca. 1950s–1980s." *The International Journal of African Historical Studies* 40 (1): 119–51.

Callaci, Emily. 2011. "Dancehall Politics: Mobility, Sexuality and the Spectacles of Racial Respectability in Late Colonial Tanganyika, 1930s–1961." *The Journal of African History* 52 (3): 365–84.

Callaci, Emily. 2017. *Street Archives and City Life: Popular Intellectuals in Postcolonial Tanzania.* Duke University Press.

Carse, Ashley. 2014. *Beyond the Big Ditch: Politics, Ecology, and Infrastructure at the Panama Canal.* MIT Press.

Carse, Ashley. 2016. "Keyword: Infrastructure: How a Humble French Engineering Term Shaped the Modern World." In *Infrastructures and Social Complexity*, edited by Penny Harvey, Casper Bruun Jensen, and Atsuro Morita, 45–57. Routledge.

Carse, Ashley. 2019. "The Feel of 13,000 Containers: How Pilots Learn to Navigate Changing Logistical Environments." *Ethnos.* https://doi.org/10.1080 /00141844.2019.1697337.

Cavell, Stanley. 1969. *Must We Mean What We Say.* Cambridge University Press.

Chalfin, Brenda. 2010. *Neoliberal Frontiers: An Ethnography of Sovereignty in West Africa.* University of Chicago Press.

Chalfin, Brenda. 2014. "Public Things, Excremental Politics, and the Infrastructure of Bare Life in Ghana's City of Tema." *American Ethnologist* 41 (1): 92–109.

Chalfin, Brenda. 2017. "'Wastelandia': Infrastructure and the Commonwealth of Waste in Urban Ghana." *Ethnos* 82 (4): 648–71.

Chalfin, Brenda. 2020. "Experiments in Excreta to Energy: Sustainability Science and Bio-Necro Collaboration in Urban Ghana." *The Cambridge Journal of Anthropology* 38 (2): 88–104.

Chari, Sharad, and Katherine Verdery. 2009. "Thinking between the Posts: Postcolonialism, Postsocialism, and Ethnography after the Cold War." *Comparative Studies in Society and History* 51 (1): 6–34.

Chatterjee, Elizabeth. 2020. "The Asian Anthropocene: Electricity and Fossil Developmentalism." *The Journal of Asian Studies* 79 (1): 3–24.

Chipole Convent. 2021. "Tulila Hydroelectric Power Station." Accessed November 14. https://chipole.org/enterprises/tulila-hydropower/.

Cohen, Adrienne. 2021. *Infinite Repertoire: On Dance and Urban Possibility in Postsocialist Guinea.* University of Chicago Press.

Coleman, Leo. 2017. *A Moral Technology: Electrification as Political Ritual in New Delhi.* Cornell University Press.

Collier, Stephen J. 2011. *Post-Soviet Social: Neoliberalism, Social Modernity, Biopolitics.* Princeton University Press.

Collier, Stephen J., and Andrew Lakoff. 2015. "Vital Systems Security: Reflexive Biopolitics and the Government of Emergency." *Theory, Culture & Society* 32 (2): 19–51.

Connolly, William E. 2013. *The Fragility of Things: Self-Organizing Processes, Neoliberal Fantasies, and Democratic Activism*. Duke University Press.

Cooksey, Brian. 2002. "The Power and the Vainglory, Anatomy of a $100 Million Malaysian IPP." In *Ugly Malaysians? South-South Investments Abused*, edited by Jomo K. S., 47–76. Institute for Black Research.

Cooksey, Brian, and Tim Kelsall. 2011. "The Political Economy of the Investment Climate in Tanzania." Africa Power and Politics Programme.

Cross, Jamie, and Declan Murray. 2018. "The Afterlives of Solar Power: Waste and Repair off the Grid in Kenya." *Energy Research & Social Science* 44:100–109.

Curinga, Matthew X. 2016. "The MOOC and the Multitude." *Educational Theory* 66 (3): 369–87.

Daniel, John, Varusha Naidoo, and Sanusha Naidu. 2003. "The South Africans Have Arrived: Post-Apartheid Corporate Expansion into Africa." In *State of the Nation: South Africa 2003-2004*, edited by John Daniel, Adam Habib, and Roger Southall, 368–90. HSRC Press.

Das, Veena. 1998. "Wittgenstein and Anthropology." *Annual Review of Anthropology* 27 (1): 171–95.

Das, Veena. 2014. "Action, Expression, and Everyday Life: Recounting Household Events." In *The Ground Between: Anthropologists Engage Philosophy*, edited by Veena Das, Michael D. Jackson, Arthur Kleinman, and Bhrigupati Singh, 279–306. Duke University Press.

Das, Veena. 2015. "What Does Ordinary Ethics Look Like?" In *Four Lectures on Ethics: Anthropological Perspectives*, 53–126. Hau Books.

Das, Veena. n.d. "Concepts Criss-Crossing." Unpublished. Accessed April 25, 2021. https://www.academia.edu/38236779/Concepts-Criss-crossing.pdf.

Das, Veena, Michael D. Jackson, Arthur Kleinman, and Bhrigupati Singh. 2014. "Introduction: Experiments between Anthropology and Philosophy; Affinities and Antagonisms." In *The Ground Between: Anthropologists Engage Philosophy*, edited by Veena Das, Michael D. Jackson, Arthur Kleinman, and Bhrigupati Singh, 1–26. Duke University Press.

Dean, Erin. 2020. "Uneasy Entanglements: Solar Energy Development in Zanzibar." *The Cambridge Journal of Anthropology* 38 (2): 53–70.

De Boeck, Filip. 1998. "Domesticating Diamonds and Dollars: Identity, Expenditure and Sharing in Southwestern Zaire (1984-1997)." *Development and Change* 29 (4): 777–810.

De Boeck, Filip. 2005. "The Apocalyptic Interlude: Revealing Death in Kinshasa." *African Studies Review* 48 (2): 11–32.

De Boeck, Filip, and Sammy Baloji. 2016. *Suturing the City: Living Together in Congo's Urban Worlds*. Autograph ABP.

De Boeck, Filip, and Marie Francois Plissart. 2004. *Kinshasa: Tales of the Invisible City*. Ludion/Royal Museum for Central Africa.

Degani, Michael. 2017. "La Véranda, le Climatiseur et la Centrale Électrique: Race et Électricité Tanzanie Postsocialiste." *Afrique Contemporaine* 261–62 (1): 103–18.

Degani, Michael. 2018. "Shock Humor: Zaniness and the Freedom of Permanent Improvisation in Urban Tanzania." *Cultural Anthropology* 33 (3): 473–98.

Degani, Michael. 2021. "Cutting without Cutting Connection: The Semiotics of Power Patrols in Urban Tanzania." *Signs and Society* 9 (2): 176–203.

Degani, Michael, Brenda Chalfin, and Jamie Cross. 2020. "Introduction: Fuelling Capture: Africa's Energy Frontiers." *The Cambridge Journal of Anthropology* 38 (2): 1–18.

De Laet, Marianne, and Annemarie Mol. 2000. "The Zimbabwe Bush Pump: Mechanics of a Fluid Technology." *Social Studies of Science* 30 (2): 225–63.

DeLanda, Manuel. 2016. *Assemblage Theory.* Edinburgh University Press.

Delehanty, Sean. 2020. "From Modernization to Villagization: The World Bank and Ujamaa." *Diplomatic History* 44 (2): 289–314.

Deleuze, Gilles, and Félix Guattari. 1987. *A Thousand Plateaus: Capitalism and Schizophrenia.* Translated by Brian Massumi. University of Minnesota Press.

Doherty, Jacob. 2017. "Life (and Limb) in the Fast-Lane: Disposable People as Infrastructure in Kampala's Boda Industry." *Critical African Studies* 9 (2): 192–209.

Dua, Jatin. 2019. *Captured at Sea: Piracy and Protection in the Indian Ocean.* University of California Press.

Dunn, Elizabeth C. 2005. "Standards and Person-Making in East Central Europe." In *Global Assemblages: Technology, Politics, and Ethics as Anthropological Problems,* edited by Aiwha Ong and Stephen J. Collier, 173–93. Wiley.

Durkheim, Emile. (1914) 2005. "The Dualism of Human Nature and its Social Conditions." *Durkheimian Studies* 11 (1): 35–45.

Dye, Barnaby, and Joerg Hartmann. 2017. "True Cost of Power: The Facts and Risks of Building Stiegler's Gorge Hydropower Dam in Selous Game Reserve, Tanzania." World Wildlife Federation International.

Dzansi, James, Steven L. Puller, Brittany Street, and Belinda Yebuah-Dwamena. 2018. *The Vicious Circle of Blackouts and Revenue Collection in Developing Economies: Evidence from Ghana.* International Growth Centre Working Paper E-89457-GHA-1.

Eberhard, Anton, and Katharine Gratwick. 2011. "When the Power Comes: An Analysis of IPPs in Africa." Infrastructure Consortium on Africa.

Eberhard, Anton, Katharine Gratwick, and Laban Kariuki. 2018. "A Review of Private Investment in Tanzania's Power Generation Sector." *Journal of Energy in Southern Africa* 29 (2): 1–11.

Edwards, Paul N. 2003. "Infrastructure and Modernity: Force, Time, and Social Organization in the History of Sociotechnical Systems." In *Modernity and Technology,* edited by Andrew Feenberg, Philip Brey, and Thomas J. Misa, 185–225. MIT Press.

Eglash, Ron, and Ellen K. Foster. 2017. "On the Politics of Generative Justice: African Traditions and Maker Communities." In *What Do Science, Technology, and*

Innovation Mean from Africa?, edited by Clapperton Chakanetsa Mavhunga, 117–36. MIT Press.

Ellis, Stephen. 1999. *The Mask of Anarchy: The Destruction of Liberia and the Religious Dimension of an African Civil War*. Hurst.

Elyachar, J. 2010. "Phatic Labor, Infrastructure, and the Question of Empowerment in Cairo." *American Ethnologist* 37 (3): 452–64.

Esposito, Roberto, and Zakiya Hanafi. 2013. "Community, Immunity, Biopolitics." *Angelaki* 18 (3): 83–90.

Evans-Pritchard, E. E. 1937. *Witchcraft, Magic, and Oracles among the Azande*. Clarendon.

Evans-Pritchard, E. E. (1940) 2011. *The Nuer: A Description of the Modes of Livelihood and Political Institutions of a Nilotic People*. Nabu Press.

Feierman, Steven M. 1990. *Peasant Intellectuals: Anthropology and History in Tanzania*. University of Wisconsin Press.

Fennell, Catherine. 2015. *Last Project Standing: Civics and Sympathy in Post-Welfare Chicago*. University of Minnesota Press.

Ferguson, James. 1985. "The Bovine Mystique: Power, Property and Livestock in Rural Lesotho." *Man* 20 (4): 647–74.

Ferguson, James. 1994. *The Anti-Politics Machine: Development," Depoliticization, and Bureaucratic Power in Lesotho*. University of Minnesota Press.

Ferguson, James. 1999. *Expectations of Modernity: Myths and Meanings of Urban Life on the Zambian Copperbelt*. University of California Press.

Ferguson, James. 2005a. "Decomposing Modernity: History and Hierarchy after Development." *Postcolonial Studies and Beyond*, 166–81.

Ferguson, James. 2005b. "Seeing Like an Oil Company: Space, Security, and Global Capital in Neoliberal Africa." *American Anthropologist* 107 (3): 377–82.

Ferguson, James. 2006. *Global Shadows: Africa in the Neoliberal World Order*. Duke University Press.

Ferguson, James. 2013. "Declarations of Dependence: Labour, Personhood, and Welfare in Southern Africa." *Journal of the Royal Anthropological Institute* 19 (2): 223–42.

Ferguson, James. 2015. *Give a Man a Fish: Reflections on the New Politics of Distribution*. Duke University Press.

Ferguson, James. 2021 *Presence and Social Obligation: An Essay on the Share*. Chicago: Prickly Paradigm.

Fioratta, Susanna. 2015. "Beyond Remittance: Evading Uselessness and Seeking Personhood in Fouta Djallon, Guinea." *American Ethnologist* 42 (2): 295–308.

Fortes, Meyer. 1959. *Oedipus and Job in West African Religion*. Cambridge University Press.

Fouéré, Marie-Aude. 2014. "Julius Nyerere, Ujamaa, and Political Morality in Contemporary Tanzania." *African Studies Review* 57 (1): 1–24.

Fukuyama, Francis. 1989. "The End of History?" *National Interest*, no. 16: 3–18.

Galloway, Alexander R. 2013. "Love of the Middle." In *Excommunication: Three Inquiries in Media and Mediation*, edited by Eugene Thacker, McKenzie Wark, and Alexander R. Galloway, 25–76. University of Chicago Press.

Gastrow, Claudia. 2017. "Cement Citizens: Housing, Demolition and Political Belonging in Luanda, Angola." *Citizenship Studies* 21 (2): 224–39.

Geertz, Clifford. 1980. *Negara: The Theatre State in 19th Century Bali*. Princeton University Press.

"Generating Power and Cash." 2011. *Africa Confidential* 52 (12): 9.

Gershon, Ilana. 2010. "Breaking Up Is Hard to Do: Media Switching and Media Ideologies." *Journal of Linguistic Anthropology* 20 (2): 389–405.

Ghanadan, Rebecca. 2008. "Public Service or Commodity Goods? Electricity Reforms, Access and the Politics of Development in Tanzania." PhD dissertation, University of California–Berkeley.

Ghanadan, Rebecca. 2009. "Connected Geographies and Struggles over Access: Electricity Commercialization in Tanzania." In *Electric Capitalism: Recolonizing Africa on the Power Grid*, edited by David A. McDonald, 400–436. Kapstadt.

Ghanadan, Rebecca, and Anton Eberhard. 2007. "Electrical Utility Management Contracts in Africa: Lessons and Experience from the TANESCO-NETGroup Solutions Management Contract in Tanzania, 2002–2006." *MIR Working Paper*.

Glassman, Jonathan. 1991. "The Bondsman's New Clothes: The Contradictory Consciousness of Slave Resistance on the Swahili Coast." *The Journal of African History* 32 (2): 277–312.

Glassman, Jonathan. 1995. *Feasts and Riot: Revelry, Rebellion, and Popular Consciousness on the Swahili Coast, 1856–1888*. Heinemann.

Gledhill, John. 2018. "Neoliberalism." In *The International Encyclopedia of Anthropology*, edited by H. Callan. https://doi.org/10.1002/9781118924396.wbiea2073.

Glendinning, Victoria. 2006. *Electricity*. Pocket Books.

Gluckman, Max. 1968. "Inter-Hierarchical Roles: Professional and Party Ethics in Tribal Areas in South and Central Africa." In *Local Level Politics*, edited by Marc J. Swartz, 69–94. Aldine.

Godhino, Catarina, and Anton Eberhard. 2018. "Power Sector Reform and Regulation in Tanzania." In *Tanzania Institutional Diagnostic*, edited by François Bourguignon and Samuel Wangwe, 1–51. Economic Development & Institutions. Accessed April 26, 2022. https://edi.opml.co.uk/wpcms/wp-content/uploads/2018/09/07-TID_Power-sector.pdf.

Goody, Jack, ed. 1958. *The Developmental Cycle in Domestic Groups*. Cambridge University Press.

Graeber, David. 2001. *Toward an Anthropological Theory of Value: The False Coin of our Own Dreams*. Springer.

Graham, Stephen, and Simon Marvin. 2001. *Splintering Urbanism: Networked Infrastructures, Technological Mobilities and the Urban Condition*. Psychology Press.

Gratwick, Katharine, and Anton Eberhard. 2008. "Demise of the Standard Model for Power Sector Reform and the Emergence of Hybrid Power Markets." *Energy Policy* 36 (10): 3948–60.

Gratwick, Katharine, Rebecca Ghanadan, and Anton Eberhard. 2006. "Generating Power and Controversy: Understanding Tanzania's Independent Power Projects." *Journal of Energy in Southern Africa* 17 (4): 39–56.

Gray, Hazel S. 2015. "The Political Economy of Grand Corruption in Tanzania." *African Affairs* 114 (456): 382–403.

Green, Maia. 2010. "After Ujamaa? Cultures of Governance and the Representation of Power in Tanzania." *Social Analysis* 54 (1): 15–34.

Grundy, Trevor. 2017. "Frene Ginwala, the Lenin Supplement, and the Storm Drains of History." Politicsweb, August 15. https://www.politicsweb.co.za /opinion/frene-ginwala-the-lenin-supplement-and-the-storm-d.

Günel, Göcke. 2019. *Spaceship in the Desert: Energy, Climate Change, and Urban Design in Abu Dhabi.* Duke University Press.

Guyer, Jane. 1993a. "Toiling Ingenuity: Food Regulation in Britain and Nigeria." *American Ethnologist* 20 (4): 797–817.

Guyer, Jane. 1993b. "Wealth in People and Self-Realization in Equatorial Africa." *Man* 28 (2): 243–65.

Guyer, Jane. 1997. "Endowments and Assets: The Anthropology of Wealth and the Economics of Intrahousehold Allocation." In *Intrahousehold Resource Allocation in Developing Countries: Methods, Models, and Policy,* edited by Lawrence Haddad, John Hoddinott, and Harold Alderman, 112–28. Johns Hopkins University Press.

Guyer, Jane. 2004. *Marginal Gains: Monetary Transactions in Atlantic Africa.* University of Chicago Press.

Guyer, Jane. 2007. "Prophecy and the Near Future: Thoughts on Macroeconomic, Evangelical, and Punctuated Time." *American Ethnologist* 34 (3): 409–21.

Guyer, Jane. 2010. "The Eruption of Tradition?" *Anthropological Theory* 10 (1–2): 123–31.

Guyer, Jane, and Samuel M. Eno Belinga. 1995. "Wealth in People as Wealth in Knowledge: Accumulation and Composition in Equatorial Africa." *Journal of African History* 36 (1): 91–120.

Hage, Ghassan. 2013. "Eavesdropping on Bourdieu's Philosophers." *Thesis Eleven* 114 (1): 76–93.

Han, Clara. 2012. *Life in Debt: Times of Care and Violence in Neoliberal Chile.* University of California Press.

Harbach, Chad. 2008. "The End." *N+1,* Winter. http://nplusonemag.com/the-end -the-end-the-end.

Harman, Graham. 2010. *Prince of Networks: Bruno Latour and Metaphysics.* Re press.

Harms, Erik. 2011. *Saigon's Edge: On the Margins of Ho Chi Minh City.* University of Minnesota Press.

Harris, Clive. 2003. *Private Participation in Infrastructure in Developing Countries: Trends, Impacts, and Policy Lessons.* World Bank.

Harrison, Faye, Carole McGranahan, Kaifa Roland, and Bianca C. Williams. 2016. "Decolonizing Anthropology: A Conversation with Faye Harrison Part I." *Anthrodendum* (blog), May 2. https://savageminds.org/2016/05/02 /decolonizing-anthropology-a-conversation-with-faye-v-harrison-part-i/.

Hart, Keith. 1973. "Informal Income Opportunities and Urban Employment in Ghana." *The Journal of Modern African Studies* 11 (1): 61–89.

Hart, Keith. 1982. *The Political Economy of West African Agriculture*. Cambridge University Press.

Hart, Keith. 1986. "Heads or Tails? Two Sides of the Coin." *Man* 21 (4): 637–56.

Hart, Keith. 2000. "Kinship, Contract, and Trust: The Economic Organization of Migrants in an African City Slum." In *Trust: Making and Breaking Cooperative Relations*, edited by Diego Gambetta, 176–93. Blackwell.

Hart, Keith. 2004. "From Bell Curve to Power Law: Distributional Models between National and World Society." *Social Analysis* 48 (3): 220–24.

Hart, Keith. 2005. "Notes towards an Anthropology of Money." *Kritikos* 2. Accessed April 26, 2022. https://intertheory.org/hart.htm.

Hart, Keith. 2006. "Bureaucratic Form and the Informal Economy." In *Linking the Formal and Informal Economies: Examples from Developing Countries*, edited by Basudeb Guha-Khasnobis, Ravi Kanbur, and Elinor Ostrom, 21–35. Oxford University Press.

Harvey, David. 2007. *A Brief History of Neoliberalism*. Oxford University Press.

Harvey, David. 2012. *Rebel Cities: From the Right to the City to the Urban Revolution*. Verso.

Hasenöhrl, Ute. 2018. "Rural Electrification in the British Empire." *History of Retailing and Consumption* 4 (1): 10–27.

Hay, Winston. 1998. "Tanzania: Power Loss Reduction Study. Volume 2: Transmission and Distribution System Reduction of Non-Technical Losses." UNDP/World Bank Energy Sector Management Assistance Program Report 204B/98.

Heidenreich, Elisabeth. 2009. "Spaces of Flow as Technical and Cultural Mediators between Society and Nature." *Environment, Development & Sustainability* 11 (6): 1145–54.

Heilman, Bruce. 1998. "Who Are the Indigenous Tanzanians? Competing Conceptions of Tanzanian Citizenship in the Business Community." *Africa Today* 45 (3–4): 369–87.

Hénaff, Marcel. 2013. "Living with Others: Reciprocity and Alterity in Lévi-Strauss." *Yale French Studies*, no. 123: 63–82.

Herzfeld, Michael. 2005. *Cultural Intimacy: Social Poetics of the Nation-State*. Routledge.

Hickel, Jason. 2014. "Xenophobia in South Africa: Order, Chaos, and the Moral Economy of Witchcraft." *Cultural Anthropology* 29 (1): 103–27.

High, Mette M., and Jessica M. Smith. 2019. "Introduction: The Ethical Constitution of Energy Dilemmas." *Journal of the Royal Anthropological Institute* 25 (1): 9–28.

Hilgers, Mathieu. 2011. "The Three Anthropological Approaches to Neoliberalism." *International Social Science Journal* 61 (202): 351–64.

Hilgers, Mathieu. 2012. "The Historicity of the Neoliberal State." *Social Anthropology* 20 (1): 80–94.

Hillewaert, Sarah. 2016. "'Whoever Leaves Their Traditions Is a Slave': Contemporary Notions of Servitude in an East African Town." *Africa* 86 (3): 425–46.

Hirschman, Albert O. 1970. *Exit, Voice, and Loyalty: Responses to Decline in Firms, Organizations, and States*. Harvard University Press.

Hischler, Kurt, and Rolf Hofmeier. 2009. "Tanzania." In *Africa Yearbook*, vol. 5, *Politics, Economy and Society South of the Sahara in 2008*, edited by Andreas Mehler, Henning Melber, and Klaas Van Walraven, 375–86. Brill.

Ho, Karen. 2009. *Liquidated: An Ethnography of Wall Street*. Duke University Press.

Hoag, Heather J., and May-Britt B. Öhman. 2008. "Turning Water into Power: Debates over the Development of Tanzania's Rufiji River Basin, 1945–1985." *Technology and Culture* 49 (3): 624–51.

Hobsbawm, Eric. 1969. *Bandits*. Weidenfeld & Nicolson.

Hobsbawm, Eric. 1984. "Artisan or Labour Aristocrat?" *The Economic History Review* 37 (3): 355–72.

Hoffman, Danny. 2011. *The War Machines: Young Men and Violence in Sierra Leone and Liberia*. Duke University Press.

Holterman, Devin. 2014. "Slow Violence, Extraction and Human Rights Defense in Tanzania: Notes from the Field." *Resources Policy* 40:59–65.

Hojer, Lars, and Morten Axel Pedersen. 2019. *Urban Hunters: Dealing and Dreaming in Times of Transition*. Yale University Press.

Howe, Cymene. 2019. *Ecologics: Wind and Power in the Anthropocene*. Duke University Press.

Howes, David. 2016. "The Extended Sensorium: Introduction to the Sensory and Social Thought of François Laplantine." Series editor's preface to *The Life of the Senses: Introduction to a Modal Anthropology*, by François Laplantine, vii–xiv. Bloomsbury.

Huber, Matthew T. 2013. *Lifeblood: Oil, Freedom, and the Forces of Capital*. University of Minnesota Press.

Hughes, David McDermott. 2017. *Energy without Conscience: Oil, Climate Change, and Complicity*. Duke University Press.

Hughes, Thomas P. 1987. "The Evolution of Large Technological Systems." In *The Social Construction of Technological Systems: New Directions in the Sociology and History of Technology*, edited by Thomas Parke Hughes, Trevor J. Pinch, Wiebe E. Bijker, 51–82. MIT Press.

Hunter, Emma. 2015. "Julius Nyerere, the Arusha Declaration, and the Deep Roots of a Contemporary Political Metaphor." In *Remembering Nyerere in Tanzania: History, Memory, Legacy*, edited by Marie-Aude Fouére, 73–96. Mkuki na Nyota.

Hunter, Emma. 2017. *Political Thought and the Public Sphere in Tanzania: Freedom, Democracy and Citizenship in the Era of Decolonization*. Cambridge University Press.

Hunter, Mark. 2010. *Love in the Time of AIDS Inequality, Gender, and Rights in South Africa*. Indiana University Press.

Huntington, Samuel P. 1993. *The Third Wave: Democratization in the Late Twentieth Century*. University of Oklahoma Press.

Imami, Drini, Endrit Lami, Edvin Zhllima, Muje Gjonbalaj, and Geoffrey Pugh. 2020. "Closer to Election, More Light: Electricity Supply and Elections in a Postconflict Transition Economy." *Post-Communist Economies* 32 (3): 376–90.

Ingold, Tim. 2015. *The Life of Lines*. Routledge.

Institute for Precarious Consciousness. 2014. "Anxiety, Affective Struggle, and Precarity Consciousness-Raising." *Interface: A Journal for and about Social Movements* 2 (6): 271–300.

International Law and Policy Institute. 2014. "Political Economy Analysis of the Energy Sector in Tanzania." International Law and Policy Institute.

Isaacman, Allen. 2005. "Displaced People, Displaced Energy, and Displaced Memories: The Case of Cahora Bassa, 1970–2004." *The International Journal of African Historical Studies* 38 (2): 201–38.

Ivaska, Andrew. 2011. *Cultured States: Youth, Gender, and Modern Style in 1960s Dar es Salaam*. Duke University Press.

Jackson, Michael, and Ivan Karp, eds. 1990. *Personhood and Agency*. Uppsala University Press.

Jackson, Steven J. 2014. "Rethinking Repair." In *Media Technologies*, edited by Tarleton Gillespie, Pablo J. Boczkowski, and Kirsten A. Foot, 221–40. MIT Press.

Jackson, Steven J., Alex Pompe, and Gabriel Krieshok. 2012. "Repair Worlds: Maintenance, Repair, and ICT for Development in Rural Namibia." In *Proceedings of the ACM 2012 Conference on Computer Supported Cooperative Work*, 107–16. ACT.

Jacome, Veronica, and Isha Ray. 2018. "The Prepaid Electric Meter: Rights, Relationships and Reification in Unguja, Tanzania." *World Development* 105:262–72.

Jakobson, Roman. 1960. "Closing Statement: Linguistics and Poetics." In *Style in Language*, edited by Thomas A. Sebeok, 350–377. MIT Press.

Jennings, Michael. 2002. "'Almost an Oxfam in Itself': Oxfam, Ujamaa and Development in Tanzania." *African Affairs* 101 (405): 509–30.

Jensen, Casper Bruun, and Atsuro Morita. 2017. "Introduction: Infrastructures as Ontological Experiments." *Ethnos* 82 (4): 615–26.

John, Samwel, Gordon McGranahan, Mwanakombo Mkanga, Tim Ndezi, Stella Stephen, and Cecilia Tacoli. 2020. "The Churn of the Land Nexus and Contrasting Gentrification Processes in Dar Es Salaam and Mwanza, Tanzania." *Environment and Urbanization* 32 (2): 429–46.

Johnson, Andrew Alan. 2013. "Progress and its Ruins: Ghosts, Migrants, and the Uncanny in Thailand." *Cultural Anthropology* 28 (2): 299–319.

Kabwe, Zitto. 2014. "How PAP Acquired IPTL for Almost Nothing and Looted US$124m from the BoT." *TransparentTanzania* (blog), August 24. https://escrowscandaltz.wordpress.com/2014/08/24/how-pap-acquired-iptl-for-almost-nothing-and-looted-us124m-from-the-bot/.

Kaika, Maria, and Erik Swyngedouw. 2000. "Fetishizing the Modern City: The Phantasmagoria of Urban Technological Networks." *International Journal of Urban and Regional Research* 24 (1): 120–38.

Kapferer, Bruce. 2005. "New Formations of Power, the Oligarchic-Corporate State, and Anthropological Ideological Discourse." *Anthropological Theory* 5 (3): 285–99.

Karp, Ivan. 2002. "Development and Personhood." In *Critically Modern Alternatives, Alterities, Anthropologies*, edited by Bruce M. Knauft, 62–104. Indiana University Press.

Kasoga, Lenny B. 1998. "Privatization and Government Regulation of Public Utilities: Just and Reasonable Return—The Case of TANESCO." *African Journal of Finance and Management* 7 (1): 33–38.

Kelsall, Tim. 2002. "Shop Windows and Smoke-Filled Rooms: Governance and the Re-Politicisation of Tanzania." *The Journal of Modern African Studies* 40 (4): 597–619.

Kerner, Donna O. 2019. "'Hard Work' and Informal Sector Trade in Tanzania." In *Traders Versus the State*, edited by Gracia Clark, 41–56. Routledge.

Khan, Mushtaq H. 2018. "Political Settlements and the Analysis of Institutions." *African Affairs* 117 (469): 636–55.

Khan, Naveeda. 2006. "Flaws in the Flow: Roads and Their Modernity in Pakistan." *Social Text* 24 (4): 87–113.

Khan, Naveeda. 2019. "At Play with the Giants: Between the Patchy Anthropocene and Romantic Geology." *Current Anthropology* 60 (S20): S333–41.

Kiangi, Ludia Stanley. 2015. "Assessing Customer Satisfaction on Electricity Conventional Billing System in Tanzania: The Case of Tanesco Ilala Region." PhD dissertation, Open University of Tanzania.

Klein, Naomi. 2007. *The Shock Doctrine: The Rise of Disaster Capitalism*. Macmillan.

Kockelman, Paul. 2016. "Grading, Gradients, Degradation, Grace Part 2: Phenomenology, Materiality, and Cosmology." *HAU: Journal of Ethnographic Theory* 6 (3): 337–65.

Kockelman, Paul. 2017. *The Art of Interpretation in the Age of Computation*. Oxford University Press.

Kohn, Eduardo. 2013. *How Forests Think: Toward an Anthropology Beyond the Human*. University of California Press.

Kombe, Wilbard Jackson. 2005. "Land Use Dynamics in Peri-Urban Areas and Their Implications on the Urban Growth and Form: The Case of Dar Es Salaam, Tanzania." *Habitat International* 29 (1): 113–35.

Kopytoff, Igor. 1986. "The Cultural Biography of Things: Commoditization as Process." In *The Social Life of Things: Commodities in Cultural Perspective*, edited by Arjun Appadurai, 64–94. Oxford University Press.

Kornai, János. 1986. "The Soft Budget Constraint." *Kyklos* 39 (1): 3–30.

Kuper, Adam. 1970. "Gluckman's Village Headman." *American Anthropologist* 72 (2): 355–58.

Kusimba, Sibel. 2020. "Embodied Value: Wealth-in-People." *Economic Anthropology* 7 (2): 166–75.

Kyando, Moses, Dennis Ikanda, and Eivin Røskaft. 2017. "Hotspot Elephant-Poaching Areas in the Eastern Selous Game Reserve, Tanzania." *African Journal of Ecology* 55 (3): 365–71.

Lacan, Jacques. (1966) 2007. *Écrits: The First Complete Edition in English*. Translated by Bruce Fink. Norton.

Lal, Priya. 2015. *African Socialism in Postcolonial Tanzania: Between the Village and the World*. Cambridge University Press.

Langwick, Stacey A. 2007. "Devils, Parasites, and Fierce Needles: Healing and the Politics of Translation in Southern Tanzania." *Science, Technology & Human Values* 32 (1): 88–117.

Laplantine, François. 2016. *The Life of the Senses: Introduction to a Modal Anthropology*. Bloomsbury.

Larkin, Brian. 2008. *Signal and Noise Media, Infrastructure, and Urban Culture in Nigeria*. Duke University Press.

Larkin, Brian. 2013. "The Politics and Poetics of Infrastructure." *Annual Review of Anthropology* 42 (1): 327–43.

Larkin, Brian. 2018. "Promising Forms: The Political Aesthetics of Infrastructure." In *The Promise of Infrastructure*, edited by Nikhil Anand, Akhil Gupta, and Hannah Appel, 175–202. Duke University Press.

Latour, Bruno. 1987. "The Enlightenment without the Critique: A Word on Michel Serres' Philosophy." *Royal Institute of Philosophy Supplements* 21:83–97.

Latour, Bruno. 2005. *Reassembling the Social: An Introduction to Actor-Network-Theory*. Oxford University Press.

Latour, Bruno. 2012. "The Modes of Existence Project: An Exercise in Collective Inquiry and Digital Humanities." Lecture at University of Cambridge, November 6. Accessed August 29, 2021. https://youtu.be/gL3WBHTWDjI.

Le Corbusier. (1923) 2013. *Towards a New Architecture*. Dover Publications.

Lemon, Alaina. 2018. *Technologies for Intuition: Cold War Circles and Telepathic Rays*. University of California Press.

Lennart, Larsson, Thomas Davy, Jönsson Mikael, and Lars Hagström. 2014. "Joint Energy Sector Review 2012/2013." Tanzania Ministry of Energy and Minerals/Swedish International Development Cooperation Agency.

Lévi-Strauss, Claude. 1966. *The Savage Mind*. University of Chicago Press.

Lévi-Strauss, Claude. 1969. *The Elementary Structures of Kinship*. Beacon Press.

Lewinson, Anne. 2003. "Imagining the Metropolis, Globalizing the Nation: Dar Es Salaam and National Culture in Tanzanian Cartoons." *City & Society* 15 (1): 9–30.

Lewinson, Anne. 2006a. "Domestic Realms, Social Bonds, and Class: Ideologies and Indigenizing Modernity in Dar Es Salaam, Tanzania." *Canadian Journal of African Studies/La Revue Canadienne Des Études Africaines* 40 (3): 462–95.

Lewinson, Anne. 2006b. "Love in the City: Navigating Multiple Relationships in Dar Es Salaam, Tanzania." *City & Society* 18 (1): 90–115.

Littlechild, Stephen, and Lynne Kiesling. 2021. "Hayek and the Texas Blackout." *The Electricity Journal* 34 (6): 1–9. https://doi.org/10.1016/j.tej.2021.106969.

Lofchie, Michael F. 2014. *The Political Economy of Tanzania*. University of Pennsylvania Press.

Lombard, Louisa. 2013. "Navigational Tools for Central African Roadblocks." *PoLAR: Political and Legal Anthropology Review* 36 (1): 157–73.

Lombard, Louisa. 2016. "Threat Economies and Armed Conservation in Northeastern Central African Republic." *Geoforum* 69:218–26.

Lukumai, Emmanuel. 2006. "The Implementation of Civil Service Reforms in Tanzania, 1991–2000." Master's thesis, University of Bergen.

Lyons, Kristin. 2020. *Vital Decomposition: Soil Practitioners and Life Politics*. Duke University Press.

Maddox, Gregory, and James Leonard Giblin, eds. 2005. *In Search of a Nation: Histories of Authority & Dissidence in Tanzania*. James Currey Ltd.

Mains, Daniel. 2019. *Under Construction: Technologies of Development in Urban Ethiopia*. Duke University Press.

Malinowski, Bronislaw. (1922) 2002. *Argonauts of the Western Pacific: An Account of Native Enterprise and Adventure in the Archipelagoes of Melanesian New Guinea*. Routledge.

Malinowski, Bronislaw. (1923) 1972. "Phatic Communion." In *Communication in Face to Face Interaction*, edited by John Laver and Sandy Hutcheson, 146–52. Penguin.

Mampilly, Zachariah. 2013. "Accursed by Man, Not God: The Fight for Tanzania's Gas Lands." Warscapes, August 20. http://www.warscapes.com/reportage /accursed-man-not-god-fight-tanzanias-gas-lands.

Marcus, George E. 1995. "Ethnography in/of the World System: The Emergence of Multi-Sited Ethnography." *Annual Review of Anthropology* 24:95–117.

Marx, Karl. (1852) 2008. *The 18th Brumaire of Louis Bonaparte*. Wildside Press.

Massumi, Brian. 2017. *The Principle of Unrest: Activist Philosophy in the Expanded Field*. University of Minnesota Press.

Mathews, Gordon, and Yang Yang. 2012. "How Africans Pursue Low-End Globalization in Hong Kong and Mainland China." *Journal of Current Chinese Affairs* 41 (2): 95–120.

Matiku, Emmanuel, Andrew Mbwambo, and Joseph Kimeme. 2011. "Rent-Seeking Behaviour and the Organisation of Utilities." In *Theories and Stories in African Public Administration*, edited by Josephat Itika, Ko de Ridder, and Albertian Tollenaa, 189–203. Leiden African Studies Centre.

Maurer, Bill. 2012. "Payment: Forms and Functions of Value Transfer in Contemporary Society." *Cambridge Anthropology* 30 (2): 15–35.

Mauss, Marcel. (1925) 2016. *The Gift: Expanded Edition*. Translated by Jane Guyer. HAU Books.

Mauss, Marcel. 1973. "Techniques of the Body." *Economy and Society* 2 (1): 70–88.

Mazrui, Ali. 1967. "Tanzaphilia." *Transition*, no. 31: 20–26.

Mazzarella, William. 2017. *The Mana of Mass Society*. University of Chicago Press.

Mbembe, Achille. 2001. *On the Postcolony*. University of California Press.

Mbembe, Achille. 2021. *Out of the Dark Night: Essays on Decolonization*. Columbia University Press.

Mbembe, Achille, and Sarah Nuttall. 2004. "Writing the World from an African Metropolis." *Public Culture* 16 (3): 347–72.

Mbilinyi, Tumaini. 2013. "Assessment of Revenue Collections at Government Agencies in Tanzania: The Case of Tanzania Electricity Supply Company Limited (Tanesco)." PhD dissertation, Mzumbe University.

McGovern, Mike. 1999. "Durkheim and Heidegger: Two Social Ontologies and Some Implications." *Journal of the Anthropological Society of Oxford* 29 (2): 105–20.

McGovern, Mike. 2010. *Making War in Cote d'Ivoire*. University of Chicago Press.

McGovern, Mike. 2012a. "Turning the Clock Back or Breaking with the Past? Charismatic Temporality and Elite Politics in Côte d'Ivoire and the United States." *Cultural Anthropology* 27 (2): 239–60.

McGovern, Mike. 2012b. *Unmasking the State: Making Guinea Modern*. University of Chicago Press.

McGovern, Mike. 2015. "Liberty and Moral Ambivalence: Postsocialist Transitions, Refugee Hosting, and Bodily Comportment in the Republic of Guinea." *American Ethnologist* 42 (2): 247–61.

McGovern, Mike. 2017. *A Socialist Peace? Explaining the Absence of War in an African Country*. University of Chicago Press.

McKay, Ramah. 2017. *Medicine in the Meantime: The Work of Care in Mozambique*. Duke University Press.

Melly, Caroline. 2017. *Bottleneck: Moving, Building, and Belonging in an African City*. University of Chicago Press.

Mercer, Claire. 2003. "Performing Partnership: Civil Society and the Illusions of Good Governance in Tanzania." *Political Geography* 22 (7): 741–63.

Mercer, Claire. 2020. "Boundary Work: Becoming Middle Class in Suburban Dar Es Salaam." *International Journal of Urban and Regional Research* 44 (3): 521–36.

Middleton, John. 1994. *The World of the Swahili: An African Mercantile Civilization*. Yale University Press.

Middleton, Townsend, and Jason Cons. 2014. "Coming to Terms: Reinserting Research Assistants into Ethnography's Past and Present." *Ethnography* 15 (3): 279–90.

Miers, Suzanne, and Igor Kopytoff, eds. 1979. *Slavery in Africa: Historical and Anthropological Perspectives*. University of Wisconsin Press.

Min, Brian, and Miriam Golden. 2014. "Electoral Cycles in Electricity Losses in India." *Energy Policy* 65:619–25.

Mitchell, Timothy. 2011. *Carbon Democracy: Political Power in the Age of Oil*. Verso.

Mkobya, Juma F., Oscar Kashaigili, Lusajo K. Mkwakaliku, Mbayani Y. Saruni, Omary H. Juma, Rustis S. Bernard, John I. Kabadi, et al. 2013. "Executive Summary of the Power System Master Plan 2012 Update." The United Republic of Tanzania Ministry of Energy and Minerals.

Molony, Tom. 2014. *Nyerere: The Early Years*. Boydell & Brewer Ltd.

Munn, Nancy D. (1976) 1992. *The Fame of Gawa: A Symbolic Study of Value Transformation in a Massim (Papua New Guinea) Society*. Duke University Press.

Must, Elise. 2018. "Structural Inequality, Natural Resources and Mobilization in Southern Tanzania." *African Affairs* 117 (466): 83–108.

Mutahaba, Gelase. 2005. "Pay Reform and Corruption in Tanzania's Public Service." Paper presented at *Seminar on Potential for Public Service Pay Reform to Eradicate Corruption among Public Servants in Tanzania*. Economic and Social Research Foundation, Dar es Salaam, May 26.

Mwakikagile, Godfrey. 2006. *Tanzania under Mwalimu Nyerere: Reflections on an African Statesman*. New Africa Press.

Mwinami, Stella. 2014. "An Assessment of the Effect of Workforce Diversity on Employee Performance at Tanesco." PhD dissertation, Mzumbe University.

Myhre, Knut Christian. 2017. *Returning Life: Language, Life Force and History in Kilimanjaro*. Berghahn Books.

Nagar, Richa. 1996. "The South Asian Diaspora in Tanzania: A History Retold." *Comparative Studies of South Asia, Africa and the Middle East* 16 (2): 62–80.

Neuwirth, Robert. 2011. *Stealth of Nations: The Global Rise of the Informal Economy*. Pantheon Books.

Newell, Sasha. 2006. "Estranged Belongings: A Moral Economy of Theft in Abidjan, Côte d'Ivoire." *Anthropological Theory* 6 (2): 179–203.

Newell, Sasha. 2007. "Pentecostal Witchcraft: Neoliberal Possession and Demonic Discourse in Ivoirian Pentecostal Churches." *Journal of Religion in Africa* 37 (4): 461–90.

Newell, Sasha. 2012. *The Modernity Bluff: Crime, Consumption, and Citizenship in Côte D'Ivoire*. University of Chicago Press.

Ngonyani, Deogratis. 2002. "Ujamaa [Socialism] Metaphors: The Basis of President Nyerere's Political Terminology." In *Surviving through Obliqueness; Language of Politics in Emerging Democracies*, edited by Samuel Gyasi Obeng and Beverly Hartford, 31–43. Nova Science.

Nguyen, Vinh-Kim. 2017. "Viral Speed: Infrastructure, Connectivity, Ontogeny; or, Notes on the Molecular Epidemiology of Epidemics." *Cultural Anthropology* 32 (1): 28–34.

Nordstrom, Carolyn. 2004. *Shadows of War: Violence, Power, and International Profiteering in the Twenty-First Century*. University of California Press.

Nyamnjoh, Francis B. 2002. "'A Child is One Person's Only in the Womb': Domestication, Agency and Subjectivity in the Cameroonian Grassfields." In *Postcolonial Subjectivities in Africa*, edited by Richard Werbner, 111–138. Zed Books.

Nyamnjoh, Francis B. 2017. "Incompleteness: Frontier Africa and the Currency of Conviviality." *Journal of Asian and African Studies* 52 (3): 253–70.

Nyerere, Julius K. 1968. *Ujamaa: Essays on Socialism*. Oxford University Press.

Öhman, May-Britt. 2007. "Taming Exotic Beauties: Swedish Hydro Power Constructions in Tanzania in the Era of Development Assistance, 1960s–1990s." PhD dissertation, KTH Royal Institute of Technology.

Ojambo, Robert. 2016. "The Arab Springs and the 'Walk to Work' Movement in Uganda: Contest for Political Space and Freedom." *African Journal of Education, Science and Technology* 3 (1): 27–38.

Ong, Aiwha. 2006. *Neoliberalism as Exception: Mutations in Citizenship and Sovereignty*. Duke University Press.

Oonk, Gijsbert. 2009. *The Karimjee Jivanjee Family: Merchant Princes of East Africa 1800-2000*. Pallas Publications.

Özden-Schilling, Canay. 2021. *The Current Economy: Electricity Markets and Techno-economics*. Stanford University Press.

Paget, Dan. 2021. "Tanzania: The Authoritarian Landslide." *Journal of Democracy* 32 (2): 61–76.

Pandian, Anand. 2014. "In the Event of an Anthropological Thought." In *Wording the World: Veena Das and Scenes of Inheritance*, edited by Roma Chatterjee, 258–72. Fordham University Press.

Parks, Lisa, and Rachel Thompson. 2020. "Internet Shutdown in Africa: The Slow Shutdown: Information and Internet Regulation in Tanzania From 2010 to 2018 and Impacts on Online Content Creators." *International Journal of Communication* 14:4288–308.

Parpart, Jane L. 1984. "The 'Labor Aristocracy' Debate in Africa: The Copperbelt Case, 1924–1967." *African Economic History*, no. 13: 171–91.

Parry, Jonathan, and Maurice Bloch. 1989. *Money and the Morality of Exchange*. Cambridge University Press.

Perez, APC. 2018. "Isolated Grid Optimization (IGO) Tanzania, Tanesco Isolated Diesel Mini-Grids located in Liwale, Loliondo, Kasulu, Kibondo, and Mapanda. USAID. Accessed November 13, 2021. https://pdf.usaid.gov/pdf_docs/PA00X2MJ.pdf.

Peters, John Durham. 2015. *The Marvelous Clouds: Toward a Philosophy of Elemental Media*. University of Chicago Press.

Petit, Pierre, and Georges Mulumbwa Mutambwa. 2005. "La Crise?': Lexicon and Ethos of the Second Economy in Lubumbashi." *Africa* 75 (4): 467–87.

Phillips, Kristin. 2018. *An Ethnography of Hunger: Politics, Subsistence, and the Unpredictable Grace of the Sun*. Indiana University Press.

Phillips, Kristin. 2020. "Prelude to a Grid: Energy, Gender and Labour on an Electric Frontier." *Cambridge Journal of Anthropology* 38 (2): 71–87.

Piot, Charles. 2010. *Nostalgia for the Future: West Africa after the Cold War*. University of Chicago Press.

Pitcher, M. Anne, and Kelly Michelle Askew. 2006. "African Socialisms and Postsocialisms." *Africa: The Journal of the International African Institute* 76 (1): 1–14.

Povinelli, Elizabeth. 2011a. *Economies of Abandonment: Social Belonging and Endurance in Late Liberalism*. Duke University Press.

Povinelli, Elizabeth. 2011b. "Routes/Worlds." *E-Flux Journal*, no. 27: 1–12. http://worker01.e-flux.com/pdf/article_8888244.pdf.

Rafiq, Mohamed Yunus, Hannah Wheatley, Rashid Salti, Aloisia Shemdoe, Jitihada Baraka, and Hildegalda Mushi. 2022. "'I Let Others Speak about Condoms': Muslim Religious Leaders' Selective Engagement with an NGO-Led Family Planning Project in Rural Tanzania." *Social Science & Medicine* 293:114650. https://doi.org/10.1016/j.socscimed.2021.114650.

Rahier, Nick, Gökçe Günel, and Pamila Gupta. 2017. "Our Electric Infrastructures." Theorizing the Contemporary, *Fieldsights*, December 19. https://culanth.org/fieldsights/our-electric-infrastructures.

Rankin, William J. 2009. "Infrastructure and the International Governance of Economic Development, 1950–1965." In *Internationalization of Infrastructures*, edited by Jean-François Auger, Jan Jaap Bouma, and Rolf Künneke, 61–75. Delft University of Technology.

Ravaisson, Félix. 2008. *Of Habit*. Continuum.

Richards, Paul. 1996. *Fighting for the Rain Forest: War, Youth & Resources in Sierra Leone*. International African Institute, James Currey, and Heinemann.

Rizzo, Mateo. 2002. "Being Taken for a Ride: Privatisation of the Dar es Salaam Transport System 1983–1998." *The Journal of Modern African Studies* 40 (1): 133–57.

Robbins, Joel. 2013. "Beyond the Suffering Subject: Toward an Anthropology of the Good." *Journal of the Royal Anthropological Institute* 19 (3): 447–642.

Rogers, Douglas. 2006. "How to Be a Khoziain in a Transforming State: State Formation and the Ethics of Governance in Post-Soviet Russia." *Comparative Studies in Society and History* 48 (4): 915–45.

Rogers, Douglas. 2010. "Postsocialisms Unbound: Connections, Critiques, Comparisons." *Slavic Review* 69 (1): 1–15.

Rogers, Douglas. 2014. "Energopolitical Russia: Corporation, State, and the Rise of Social and Cultural Projects." *Anthropological Quarterly* 87 (2): 431–51.

Roitman, Janet. L. 2005. *Fiscal Disobedience: An Anthropology of Economic Regulation in Central Africa*. Princeton University Press.

Rolffs, Paula, David Ockwell, and Rob Byrne 2015. "Beyond Technology and Finance: Pay-As-You-Go Sustainable Energy Access and Theories of Social Change." *Environment and Planning A* 47 (12): 2609–27.

Rose, Nikolas. 1996. "The Death of the Social? Re-Figuring the Territory of Government." *International Journal of Human Resource Management* 25 (3): 327–56.

Rupp, Stephanie. 2016. "Circuits and Currents: Dynamics of Disruption in New York City Blackouts." *Economic Anthropology* 3 (1): 106–18.

Rutherford, Danilyn. 2016. "How Structuralism Matters." *HAU: Journal of Ethnographic Theory* 6 (3): 61–77.

Sanders, Todd. 2008. "Buses in Bongoland: Seductive Analytics and the Occult." *Anthropological Theory* 8 (2): 107–32.

Schneider, Leander. 2004. "Freedom and Unfreedom in Rural Development: Julius Nyerere, Ujamaa Vijijini, and Villagization." *Canadian Journal of African Studies/La Revue Canadienne des Études Africaines* 38 (2): 344–92.

Schwenkel, Christina. 2015. "Spectacular Infrastructure and Its Breakdown in Socialist Vietnam." *American Ethnologist* 42 (3): 520–34.

Schwenkel, Christina, and Ann Marie Leshkowich. 2012. "Guest Editors' Introduction: How Is Neoliberalism Good to Think Vietnam? How Is Vietnam Good to Think Neoliberalism?" *positions: asia critique* 20 (2): 379–401.

Scott, James. C. 1977. *The Moral Economy of the Peasant: Rebellion and Subsistence in Southeast Asia*. Yale University Press.

Scott, James. C. 1985. *Weapons of the Weak: Everyday Forms of Peasant Resistance*. Yale University Press.

Scott, James C. 1990. *Domination and the Arts of Resistance: Hidden Transcripts*. Yale University Press.

Scott, James C. 1998. *Seeing like a State: How Certain Schemes to Improve the Human Condition Have Failed*. Yale University Press.

Scott, James C. 2006. "High Modernist Social Engineering: The Case of the Tennessee Valley Authority." *Experiencing the State*, edited by Lloyd I. Rudolph and John Kurt Jacobsen, 321–32. Oxford University Press.

Scott, James C. 2012. *Two Cheers for Anarchism: Six Easy Pieces on Autonomy, Dignity, and Meaningful Work and Play*. Princeton University Press.

Scotton, Carol M. M. 1965. "Some Swahili Political Words." *Journal of Modern African Studies* 3 (4): 527–41.

Serres, Michel. 1982. *The Parasite*. Translated by Lawrence R. Schehr. Johns Hopkins University Press.

Shamir, Ronen. 2018. "Head-Hunters and Knowledge-Gatherers: Colonialism, Engineers and Fields of Planning." *History and Anthropology* 29 (4): 469–92.

Shao, John. 1986 "The Villagization Program and the Disruption of the Ecological Balance in Tanzania." *Canadian Journal of African Studies/La Revue Canadienne des Études Africaines* 20 (2): 219–39.

Shibano, Kyohei, and Gento Mogi. 2020. "Electricity Consumption Forecast Model Using Household Income: Case Study in Tanzania." *Energies* 13 (10): 2497. https://doi.org/10.3390/en13102497.

Shryock, Andrew. 2019. "Keeping to Oneself: Hospitality and the Magical Hoard in the Balga of Jordan." *History and Anthropology* 30 (5): 546–62.

Simone, AbdouMaliq. 2004a. *For the City yet to Come: Changing African Life in Four Cities*. Duke University Press.

Simone, AbdouMaliq. 2004b. "People as Infrastructure: Intersecting Fragments in Johannesburg." *Public Culture* 16 (3): 407–29.

Singh, Bhrigupati. 2014. "How Concepts Make the World Look Different: Affirmative and Negative Genealogies of Thought." In *The Ground Between: Anthropologists Engage Philosophy*, edited by Veena Das, Michael D. Jackson, Arthur Kleinman, and Bhrigupati Singh. Durham, 159–86. Duke University Press.

Sitta, Samuel, Willibrod P. Slaa, and John Cheyo. 2008. *Bunge Lenye Meno: A Parliament with Teeth for Tanzania*. Africa Research Institute.

Smiley, Sarah L. 2009. "The City of Three Colors: Segregation in Colonial Dar Es Salaam, 1891–1961." *Historical Geography* 37 (1): 178–96.

Spence, Lester K. 2015. *Knocking the Hustle: Against the Neoliberal Turn in Black Politics*. Punctum Books.

Star, Susan Leigh. 1999. "The Ethnography of Infrastructure." *American Behavioral Scientist* 43 (3): 377–91.

Stark, Laura. 2017. "Cultural Politics of Love and Provision among Poor Youth in Urban Tanzania." *Ethnos* 82 (3): 569–91.

Stevenson, Lisa. 2017. "Sounding Death, Saying Something." *Social Text* 35 (1): 59–78.

Strathern, Marilyn. 1996. "Cutting the Network." *Journal of the Royal Anthropological Institute* 2 (3): 517–35.

Stroeken, Koen. 2005. "Immunizing Strategies: Hip-Hop and Critique in Tanzania." *Africa* 75 (4): 488–509.

Takabvirwa, Kathryn. 2018. "On the Threshing Floor: Roadblocks and the Policing of Everyday Life in Zimbabwe." PhD dissertation, Stanford University.

Terray, Emmanuel. 1986. "Le Climatiseur et la Véranda." In *Afrique Plurielle, Afrique Actuelle: Hommage à Georges Balandier*, edited by Georges Balandier, 37–44. Karthala.

Thaler, Richard H., and Cass R. Sunstein. 2008. *Nudge: Improving Decisions Using the Architecture of Choice*. Yale University Press.

Thompson, Edward. P. 1971. "The Moral Economy of the English Crowd in the Eighteenth Century." *Past and Present*, no. 50: 76–136.

Tischler, Julia. 2014. "Cementing Uneven Development: The Central African Federation and the Kariba Dam Scheme." *Journal of Southern African Studies* 40 (5): 1047–64.

Tripp, Aili Marie. 1997. *Changing the Rules: The Politics of Liberalization and the Urban Informal Economy in Tanzania*. University of California Press.

Trovalla, Eric, and Ulrika Trovalla. 2015. "Infrastructure as a Divination Tool: Whispers from the Grids in a Nigerian City." *City* 19 (2–3): 332–43.

Tsing, Anna Lowenhaupt. 2000. "Inside the Economy of Appearances." *Public Culture* 12 (1): 115–44.

Tsing, Anna Lowenhaupt. 2015. *The Mushroom at the End of the World: On the Possibility of Life in Capitalist Ruins*. Princeton University Press.

Tsuruta, Tadasu. 2003. "Popular Music, Sports, and Politics: A Development of Urban Cultural Movements in Dar es Salaam, 1930s–1960s." *African Study Monographs* 24 (3): 195–222.

Turner, Victor. 1967. *Forest of Symbols: Aspects of Ndembu Ritual*. Cornell University Press.

USAID. 2018. "Hydropower in Tanzania's Rural Highlands." Accessed November 19, 2021. https://www.usaid.gov/energy/mini-grids/case-studies/tanzania-hydropower/.

Vagliasindi, Maria, and John Besant-Jones. 2013. *Power Market Structure: Revisiting Policy Options*. World Bank Publications.

Van der Straeten, Jonas 2015a. "Electrification in Tanzania from a Historical Perspective—Discourses of Development and the Marginalization of the Rural Poor." In *Micro Perspectives for Decentralized Energy Supply: Proceedings of the International Conference*, edited by Martina Schäfer, 156–61. Universitätsverlag der TU Berlin.

Van der Straeten, Jonas. 2015b. "Legacies of a Past Modernism Discourses of Development and the Shaping of Centralized Electricity Infrastructures in Late- and Postcolonial Tanzania." In *International Symposium for Next Generation Infrastructure: Conference Proceedings*, edited by Tom Dolan and Brian Collins, 275–80. UCL Department of Science, Technology, Engineering and Public Policy.

Van Huedsen, Peter. 2009. "Discipline and the New 'Logic of Delivery': Prepaid Electricity in South Africa and Beyond." In *Electric Capitalism: Recolonizing Africa on the Power Grid*, edited by David McDonald, 229–47. HSRC Press.

Venkatesan, Soumhya. 2020. "Afterword: Putting Together the Anthropology of Tax and the Anthropology of Ethics." *Social Analysis* 64 (2): 141–54.

Verdery, Katherine. 1996. *What Was Socialism, and What Comes Next?* Cambridge University Press.

Verran, Helen. 2007. "The Telling Challenge of Africa's Economies." *African Studies Review* 50 (2): 163–82.

Viveiros de Castro, Eduardo. 2014. *Cannibal Metaphysics*. University of Minnesota Press.

Vokes, Richard. 2013. "New Guinean Models in the East African Highlands." *Social Analysis* 57 (3): 95–113.

von Schnitzler, Antina. 2013. "Traveling Technologies: Infrastructure, Ethical Regimes, and the Materiality of Politics in South Africa." *Cultural Anthropology* 28 (4): 670–93.

von Schnitzler, Antina. 2016. *Democracy's Infrastructure: Techno-Politics and Protest after Apartheid*. Princeton University Press.

Wacquant, Loïc. 2012. "Three Steps to a Historical Anthropology of Actually Existing Neoliberalism." *Social Anthropology* 20 (1): 66–79.

Wallerstein, Immanuel. 1991. "Braudel on Capitalism, or Everything Upside Down." *Journal of Modern History* 63 (2): 354–61.

Walsh, Martin. 2012. "The Not-so-Great Ruaha and Hidden Histories of an Environmental Panic in Tanzania." *Journal of Eastern African Studies* 6 (2): 303–35.

Wamukonya, Njeri. 2003. "African Power Sector Reforms: Some Emerging Lessons." *Energy for Sustainable Development* 7 (1): 7–15.

Wark, McKenzie. 2014. "Furious Media: A Queer History of Heresy." In *Excommunication: Three Inquiries in Media and Mediation*, edited by Eugene Thacker, McKenzie Wark, and Alexander R. Galloway, 151–201. University of Chicago Press.

Watts, Michael. 1994. "Development II: The Privatization of Everything?" *Progress in Human Geography* 18 (3): 371–84.

Watts, Michael. 2005. "Baudelaire over Berea, Simmel over Sandton?" *Public Culture* 17 (1): 181–92.

Weber, Max. (1930) 2005. *The Protestant Ethic and the Spirit of Capitalism*. Translated by Talcott Parsons. Routledge.

Wedel, Janine R. 2009. *Shadow Elite: How the World's New Power Brokers Undermine Democracy, Government, and the Free Market*. Basic Books.

Weiss, Brad. 1996. *The Making and Unmaking of the Haya Lived World: Consumption, Commoditization, and Everyday Practice*. Duke University Press.

White, Luise. 1994. "Blood Brotherhood Revisited: Kinship, Relationship, and the Body in East and Central Africa." *Africa* 64 (3): 359–72.

White, Luise. 2000. *Speaking with Vampires: Rumor and History in Colonial Africa*. University of California Press.

Whitington, Jerome. 2019. *Anthropogenic Rivers: The Production of Uncertainty in Lao Hydropower*. Cornell University Press.

Wilder, Gary. 2014. *Freedom Time: Negritude, Decolonization, and the Future of the World*. Duke University Press.

Winther, Tanja. 2008. *The Impact of Electricity: Development, Desires and Dilemmas.* Berghahn Books.

World Bank. 1993. *The World Bank's Role in the Electric Power Sector: Policies for Effective Institutional, Regulatory, and Financial Reform.* The World Bank. https://elibrary.worldbank.org/doi/epdf/10.1596/0-8213-2318-0.

Yurchak, Alexei. 2006. *Everything Was Forever, until It Was No More: The Last Soviet Generation.* Princeton University Press.

Zárate, Salvador. 2018. "Maintenance." *Fieldsights*, March 29. https://culanth.org/fieldsights/1359-maintenance.

Zizek, Slavoj. 2005. *The Parallax View.* MIT Press.

Zizek, Slavoj. 2006. *How to Read Lacan.* Granta.

Zuckerman, Charles H. P. 2016. "Phatic Violence? Gambling and the Arts of Distraction in Laos." *Journal of Linguistic Anthropology* 26 (3): 294–314.

demand (for electricity/power), 79, 92, 112, 213n3; areas of high, 12; growing, 49, 64, 90; long-term, 51; peak, 75, 160, 195, 199

democracy, 210n12; multiparty, 48, 211n18 (*see also* multipartyism)

Democratic Republic of Congo, 8, 22, 93

deregulation, 38; media, 48; of public transportation, 139

desire, 14, 118, 131–32, 188, 203

developmentalism, 193; air-conditioner, 57; fossil, 17, 19; postcolonial, 38; socialist, 45, 147, 185; of *ujamaa*, 85

discipline, 16, 44, 188, 193; customer, 134; financial, 110; market, 48, 58; nationalist, 125; restorative, 106; socialist, 29, 189; state, 15, 113. *See also* indiscipline

Dowans Holdings, 39, 60, 65–66, 79–80, 83. *See also* Symbion Power

Durkheim, Emile, 132, 201

Eastern bloc, 10, 215n45

ecologies, 4; regional, 199

ecology, 137, 199, 204; of adaptation, 81; of circulation, 10; national, 7, 148; wet and dry season, 64

electricians, 1–2, 6, 13–14, 25, 108, 155, 158, 191; becoming infrastructure and, 165; street, 134, 147, 151, 159, 174 (see also *mafundi*); Tanesco and, 131. See also *vishoka*

electrification, 17–18, 42, 45, 140, 197; as political ritual of India, 72; street-level, 114; of Zanzibar, 13

elites, 22, 40, 49, 62, 88, 101, 154, 198, 210n12; African, 47; entrepreneurial, 34, 72; party, 19, 27, 189

emergency power, 29, 34, 59, 62–66, 69, 91, 96, 103, 148, 189; contracts, 28, 51, 88, 101; era, 75, 195; MEM and, 51; money, 67; plants, 42; scandals, 40, 89, 212n28; system of, 27; Tanesco and, 78, 80, 97, 100; World Bank funding for, 50. *See also* Richmond Development Company; Symbion Power

emergency power producers (EPPs), 32, 34–35, 40, 63, 195; arrears from, 97; contracts, 18–19, 28, 35, 37, 39, 69, 89,

91; procurement, 58. *See also* Aggreko; Symbion Power

Energy and Water Regulatory Authority (EWURA), 32–33, 66

exploitation, 15, 44, 104; Afro-socialist ideas about, 4; capitalist, 21, 40; mining and, 130; modal politics of tolerating, 61–62; political, 49; socialist definitions of, 5, 53, 203; of tenants by landlords, 122

extension (of electricity/power), 7, 26, 29, 108, 150. *See also* grid: extension of; service lines: extension of

Eyakuze, Elsie, 106–7

Ferguson, James, 8, 22, 38, 94–95, 130, 147, 209n6

form, 6, 13, 17, 50, 85–86; built, 140; of circuit, 25, 30; collective, 15; commitment to, 95; of electricity's absence, 106; emergent, 201, 205; force and, 75–76, 102, 108, 154, 175; life and, 132. *See also* social form

generation (of electricity/power), 19, 27–28, 32, 42, 83; companies, 100; crippled, 57; deficit/crisis of, 98; diesel, 51; emergency, 78; emergency power and, 40, 63, 89; facilities, 74, 103, 106; gas, 199; hydropower and, 54; IPPs and, 211n21; national, 5; network capacity for, 4; outages and, 76; as postsocialist mode, 7, 26, 72; shortfalls, 66; SPPs and, 209n3; state expansion of, 18; state subsidization of, 12; systems, 50; Tanesco and, 35, 65, 67; thermal, 27, 32, 34, 69

generators, 60–62, 65–66, 71, 80–83, 99; diesel, 17, 34, 42, 63; kerosene-powered, 52; private, 11

Ghanadan, Rebecca, 57, 217n4

gift, 100–101, 181; anthropology of, 12, 20; cutout fuse removal as, 135; exchanges, 117; metaphysics of, 11; outages and, 83; relations of infrastructure, 19; service line as, 13

Global South, 10, 110, 151; infrastructures in, 75

Swahili coast, 154–55, 185
Symbion Power, 37, 39, 65–66, 80, 97, 212n31; EPP contracts and, 35

Tabata, 105, 118
Tanganyika, 17, 21, 44
Tanganyika African National Union (TANU), 15, 44, 90, 215n42
Tennessee Valley Authority, 45, 199, 221n20
theft, 3–6, 26, 29, 112, 114, 124–25, 127, 137–38, 141, 185; culture and, 219n5; disconnections and, 111; life as, 20; meter, 175, 180; of poles, 173; retroactive, 135; temporary, 148; schemes for, 147; *vishoka* and, 163, 183
thermal power, 18, 27, 65; plants, 79; sources, 37, 42, 54
thieves, 1, 3, 137, 171, 180, 184–85, 191. See also *vishoka*
Touré, Sekou, 86, 93
tribalism (*ukabila*), 14, 130
Tungi neighborhood, 99–100, 192

Uganda, 8, 46, 90–93
ujamaa, 14, 28, 84–86, 89, 94, 210n9; policies, 211n19; villages, 44
ujanja (ingenuity), 25, 117, 146, 170, 173, 183, 189
Ulimboka, Steven, 97, 194
umma, 40, 156, 173, 175, 203. *See also* community; public good
unbundling, 50; power sector, 210n15; of Tanesco, 18, 51, 55
uswahilini neighborhoods, 29, 112, 115, 123, 125, 138–41, 147; electricity use in, 111, 117–18; metering system and, 119, 121–22.

See also Buguruni neighborhood; debt; Tabata; Vigunguti neighborhood
utumwa, 153, 161, 185, 219n4
uwazi, 46, 190. *See also* liberalization

vampires (*mumiani*), 15, 17, 125
Verran, Helen, 24, 162
VETA, 80, 158, 160, 166
vibarua (temporary workers), 113, 131, 154–55, 159–61, 164, 167, 182. See also *kibarua*
Vigunguti neighborhood, 100, 118
villagization, 45, 104, 210n11, 218n9
violence, 4, 23, 72, 74; DC teams and, 126; infrastructural, 6, 125, 129; Magufuli presidency and, 8, 194; political, 198; power interruptions and, 103; slavery and, 155; against Tanesco workers, 218n11
VIPEM Ltd., 51–52
vishoka, 3–4, 30, 113–14, 126, 133, 141, 148–52, 163–67, 171, 191; becoming infrastructure of, 155–56, 176, 182–83, 185; conmen, 143–45; meters and, 175; poles and, 173; quacks, 170; service line extension and, 181; Tanesco and, 7, 19, 26, 130–31, 134, 145, 147, 162, 193, 205; upward mobility of, 184

wealth in people, 30, 152
witchcraft, 15, 184
Wittgenstein, Ludwig, 208n16; *Philosophical Investigations*, 100
World Bank, 39–40, 45, 67, 118, 199, 210n9, 210n14; loss reduction specialists of, 128; privatization of Tanesco and, 18, 40, 50–51, 72; Songas and, 52–54; standard model of, 113; *ujamaa* policies and, 211n19

Zanzibar, 56, 192; electrification of, 13

www.ingramcontent.com/pod-product-compliance
Lightning Source LLC
Chambersburg PA
CBHW071735270326
41928CB00013B/2692